Lecture Notes in Computer Science   8978

Commenced Publication in 1973
Founding and Former Series Editors:
Gerhard Goos, Juris Hartmanis, and Jan van Leeuwen

Frank Piessens Juan Caballero
Nataliia Bielova (Eds.)

Engineering Secure Software and Systems

7th International Symposium, ESSoS 2015
Milan, Italy, March 4-6, 2015
Proceedings

 Springer

Volume Editors

Frank Piessens
KU Leuven, Department of Computer Science
Celestijnenlaan 200A
3001 Heverlee, Belgium
E-mail: frank.piessens@cs.kuleuven.be

Juan Caballero
IMDEA Software Institute
Campus de Montegancedo S/N
28223 Pozuelo de Alarcón, Spain
E-mail: juan.caballero@imdea.org

Nataliia Bielova
Inria Sophia Antipolis – Mediterranee
2004 route des Lucioles, B.P. 93
06902 Sophia Antipolis Cedex, France
E-mail: nataliia.bielova@inria.fr

ISSN 0302-9743   e-ISSN 1611-3349
ISBN 978-3-319-15617-0   e-ISBN 978-3-319-15618-7
DOI 10.1007/978-3-319-15618-7
Springer Cham Heidelberg New York Dordrecht London

Library of Congress Control Number: 2015930610

LNCS Sublibrary: SL 4 – Security and Cryptology

Typesetting: Camera-ready by author, data conversion by Scientific Publishing Services, Chennai, India

Printed on acid-free paper

Springer is part of Springer Science+Business Media (www.springer.com)

Preface

It is our pleasure to welcome you to the proceedings of the 7th International Symposium on Engineering Secure Software and Systems (ESSoS 2015). This event is part of a maturing series of symposia that attempts to bridge the gap between the software engineering and security scientific communities with the goal of supporting secure software development. The parallel technical sponsorship from ACM SIGSAC (the ACM interest group in security) and ACM SIGSOFT (the ACM interest group in software engineering) demonstrates the support from both communities and the need for providing such a bridge.

Security mechanisms and the act of software development usually go hand in hand. It is generally not enough to ensure correct functioning of the security mechanisms used. They cannot be blindly inserted into a security-critical system, but the overall system development must take security aspects into account in a coherent way. Building trustworthy components does not suffice, since the interconnections and interactions of components play a significant role in trustworthiness. Lastly, while functional requirements are generally analyzed carefully in systems development, security considerations often arise after the fact. Adding security as an afterthought, however, often leads to problems. Ad hoc development can lead to the deployment of systems that do not satisfy important security requirements. Thus, a sound methodology supporting secure systems development is needed. The presentations and associated publications at ESSoS 2015 contribute to this goal in several directions: First, improving methodologies for secure software engineering (such as formal methods and machine learning). Second, with secure software engineering results for specific application domains (such as access control, cloud, and password security). Finally, a set of presentations on security measurements and ontologies for software and systems.

The conference program featured two keynotes by Herbert Bos (Vrije Universiteit Amsterdam) and Felix Lindner (Recurity Labs GmbH), as well as research and idea papers. In response to the call for papers, 41 papers were submitted. The Program Committee selected 11 full-paper contributions (27%), presenting new research results on engineering secure software and systems. In addition, there were five idea papers, giving a concise account of new ideas in the early stages of research.

Many individuals and organizations contributed to the success of this event. First of all, we would like to express our appreciation to the authors of the submitted papers and to the Program Committee members and external referees, who provided timely and relevant reviews. Many thanks go to the Steering Committee for supporting this series of symposia, and to all the members of the Organizing Committee for their tremendous work and for excelling in their respective tasks. The DistriNet research group of the KU Leuven did an excellent job with the website and the advertising for the conference. Finally, we owe

gratitude to ACM SIGSAC/SIGSOFT, IEEE TCSP, and LNCS for continuing
to support us in this series of symposia.

December 2014   Frank Piessens
                                                      Juan Caballero
                                                      Nataliia Bielova

Organization

General Chair

Stefano Zanero   Politecnico di Milano, Italy

Program Co-chairs

Frank Piessens   Katholieke Universiteit Leuven, Belgium
Juan Caballero   IMDEA Software Institute, Spain

Publication Chair

Nataliia Bielova   Inria, France

Publicity Chair

Raoul Strackx   Katholieke Universiteit Leuven, Belgium

Web Chair

Ghita Saevels   Katholieke Universiteit Leuven, Belgium

Steering Committee

Jorge Cuellar   Siemens AG, Germany
Wouter Joosen   Katholieke Universiteit Leuven, Belgium
Fabio Massacci   Universitá di Trento, Italy
Gary McGraw   Cigital, USA
Bashar Nuseibeh   The Open University, UK
Daniel Wallach   Rice University, USA

Program Committee

Aslan Askarov   Harvard University, USA
Leyla Bilge   Symantec Research Labs, France
Stefano Calzavara   Università Ca' Foscari Venezia, Italy
Lorenzo Cavallaro   Royal Holloway, University of London, UK

Bruno Crispo	University of Trento, Italy
Werner Dietl	University of Waterloo, Canada
Michael Franz	University of California, Irvine, USA
Christian Hammer	Saarland University, Germany
Marieke Huisman	University of Twente, The Netherlands
Somesh Jha	University of Wisconsin, USA
Martin Johns	SAP Research, Germany
Christian Kreibich	Lastline, USA
Wenke Lee	Georgia Institute of Technology, USA
Zhenkai Liang	National University of Singapore, Singapore
Jay Ligatti	University of South Florida, USA
Patrick McDaniel	Pennsylvania State University, USA
Nick Nikiforakis	Stony Brook University, USA
Georgios Portokalidis	Stevens Institute of Technology, USA
Joachim Posegga	University of Passau, Germany
Alexander Pretschner	Technische Universität München, Germany
Tamara Rezk	Inria, France
Konrad Rieck	University of Göttingen, Germany
Andrei Sabelfeld	Chalmers University of Technology, Sweden
Ahmad-Reza Sadeghi	TU Darmstadt, Germany
Kapil Singh	IBM T.J. Watson Research Center, USA
Asia Slowinska	Vrije Universiteit Amsterdam, The Netherlands
Pierre-Yves Strub	IMDEA Software Institute, Spain
Carmela Troncoso	Gradiant, Spain
Xiaofeng Wang	Indiana University, USA
Mohammad Zulkernine	Queen's University, Canada

Additional Reviewers

Gunes Acar	William Harris	Davide Papini
Daniel Arp	Daniel Hedin	Juan D. Parra Rodriguez
Musard Balliu	Prachi Kumari	Silvio Ranise
Bastian Braun	Sebastian Lekies	Manuel Rudolph
Jan Cederquist	Tobias Marktscheffel	Christian Wachsmann
Drew Davidson	Dimiter Milushev	Bogdan Warinschi
Lorenzo De Carli	Martín Ochoa	Fabian Yamaguchi
Matt Fredrikson	Damien Octeau	Hong-Sheng Zhou
Alexander Fromm	Alessandro Oltramari	

Sponsoring Institutions

 POLITECNICO DI MILANO

Politecnico di Milano, Italy

NESSoS FP7 Project, Network of Excellence on Engineering Secure Future Internet Software Services and Systems, www.nessos-project.eu

PRIN Project TENACE, Protecting National Critical Infrastructures from Cyber Threats, http://www.dis.uniroma1.it/~tenace/

Table of Contents

Measurements and Ontologies

Access Control

Formal Verification of Liferay RBAC

Stefano Calzavara, Alvise Rabitti, and Michele Bugliesi

Università Ca' Foscari Venezia

Abstract. Liferay is the leading opensource portal for the enterprise, implementing a role-based access control (RBAC) mechanism for user and content management. Despite its critical importance, however, the access control system implemented in Liferay is poorly documented and lacks automated tools to assist portal administrators in configuring it correctly. To make matters worse, although strongly based on the RBAC model and named around it, the access control mechanism implemented in Liferay has a number of unconventional features, which significantly complicate verification. In this paper we introduce a formal semantics for Liferay RBAC and we propose a verification technique based on abstract model-checking, discussing sufficient conditions for the soundness and the completeness of the analysis. We then present a tool, called LIFERBAC, which implements our theory to verify the security of real Liferay portals. We show that the tool is effective at proving the absence of security flaws, while efficient enough to be of practical use.

1 Introduction

Liferay[1] is the leading opensource portal for the enterprise, adopted by important companies like Allianz, Cisco, Lufthansa and Vodafone, just to name a few [17]. Liferay allows portal administrators to conveniently manage both users and contents in a unified web framework. Users are typically structured into a hierarchy of *organizations*, where members of a child organization are also members of the parent organization. Contents, instead, are collected into *sites*, built as assemblies of different pages, portlets and social collaboration tools, like blogs and wikis. Both organizations and sites belong to a top-level *company*, and a single portal may host different companies at the same time.

For enterprises, the Liferay portal is at the core of the business process, since security-critical portlets may allow, for instance, to access sensitive information and/or to reorganize workflows. To ensure that private contents are only accessed by the intended recipients and that business processes are only handled by authorized users, Liferay implements a *role-based access control* (RBAC) mechanism.

1.1 Liferay RBAC

In the standard RBAC model, permissions are assigned to a relatively small and fixed set of roles, while roles are assigned to a potentially large and dynamic set

---

[1] http://www.liferay.com

F. Piessens et al. (Eds.): ESSoS 2015, LNCS 8978, pp. 1–16, 2015.

of users: since user privileges only depend on the assigned roles, this approach simplifies the access control management task [7]. When role administration is itself role-based, like in the case of Liferay, the RBAC model is typically called *administrative* and abbreviated as ARBAC [20]. For the sake of simplicity and for consistency with the Liferay documentation, in this paper we uniform the two models and we just use the acronym RBAC everywhere.

Though extremely popular and widely deployed, real-world RBAC systems are notoriously difficult to get right, since the set of roles dynamically assignable to each user is easily under-estimated and occasional changes to the access control policy may introduce overlooked security flaws. The research community has then proposed formal methods as an effective tool to strengthen complex RBAC systems and ensure that they meet their intended goals [3,9,8,2,19]. Notable examples of useful security goals include role (un)reachability, ensuring that a given role granting powerful privileges is never assigned to untrusted users, or mutual exclusion properties, preventing the assignment of dangerous combinations of permissions to the same user.

Despite its critical importance and these well-known problems, the access control system implemented in Liferay is poorly documented and lacks automated tools to assist portal administrators in configuring it correctly. To make matters worse, although strongly based on the RBAC model and named around it, the access control mechanism implemented in Liferay does *not* constitute, strictly speaking, an RBAC system. First, users of the portal may be allowed to *impersonate* other users and inherit all the privileges granted to them: this implies that, contrary to the RBAC model, the identity of the users is not immaterial and the verification problem becomes more challenging. Moreover, besides regular roles, Liferay also features *site roles*, *organization roles* and *owner roles*, used to constrain access rights exclusively to site members, organization members and resource owners respectively. These special roles have an unconventional semantics, reminiscent of a specific kind of *parametrized roles* [10,15,22]. Their introduction breaks a desirable property of the standard RBAC model: user privileges do not depend only on the assigned roles, but also on the state of the Liferay portal, which further complicates verification.

1.2   Contributions

Our contributions can be summarized as follows:

1. we define a formal semantics of Liferay RBAC in terms of a state transition system, which concisely and precisely captures all the subtleties of the access control model. We additionally discuss how we ensure the adequacy of the formal semantics with respect to the behaviour of the real portal;

2. we introduce an abstract semantics which provides a finite approximation of the infinite-state transition system induced by the (concrete) formal semantics. We show that the abstract semantics can be used to soundly verify useful security properties expressed in a fragment of a standard modal (temporal) logic. Moreover, we prove that, when impersonation is not used, the

adoption of the abstract semantics does not introduce any loss of precision in the security analysis;

3. we implement a tool, called LIFERBAC, which leverages the abstract semantics and the modal logic to verify the security of real Liferay portals. We show that the tool is effective at proving the absence of security flaws, while efficient enough to be of practical use on a realistic case study.

Structure of the paper. Section 2 defines the formal semantics of Liferay RBAC. Section 3 introduces the abstract semantics and studies the verification problem. Section 4 presents the tool and the experiments. Section 5 discussed related work. Section 6 concludes. The proofs of the formal results are given in the long version of the paper [4].

2   Semantics of Liferay RBAC

Liferay users are organised into a hierarchy of groups, including companies, organizations and sites. Similarly, different items in the portal are assigned to these groups on an ownership basis: for instance, a given portal page may belong to the site s, which in turn is under the control of the company c. Liferay provides various tools to grant or deny access to a given resource based on the group hierarchy: *scoping* allows to extend access rights on a group to each item belonging to that group, while *parametrized roles* like organization roles and site roles provide facilities for granting access privileges which are restricted to organization-specific/site-specific resources and organization/site members. For example, the assignment of an organization role Reader[o] may allow members of the organization o to get read access to all the items belonging to o. Finally, owner roles can be used to define access rights for individual resource owners.

2.1   Syntax

We let *Users* be an unbounded set of users and $(G, \preceq)$ be a poset of *groups*. Groups and their underlying order uniformly model different collections handled by the portal, i.e., companies, organizations and sites. We assume an unbounded set *Items*, which includes a number of resources of interest, e.g., portlets, message boards, layouts, etc. Each item i belongs to a fixed set of groups, written $i.groups$.

We assume a set of regular roles *RegRoles* and a set of *role templates* [10]. A role template $r[\cdot] \in RoleTemps$ is a role with a hole (the dot): by instantiating the hole with an object o, we generate a new *parametrized* role $r[o]$. As we formalize below, parametrized roles enforce additional runtime restrictions on the privileges which are granted by the Liferay portal. We assume that role templates are *sorted*, i.e., they are partitioned into two different sets *GrpTemps* and $\{Owner_j[\cdot]\}_{j \in J}$ (with $J = \{0, \ldots, n\}$ for some natural n) with holes of type G and *Items* respectively: we let *ParRoles* be the set of the parametrized roles obtained by instantiating the holes occurring in role templates with objects of the correct type. We let $R = RegRoles \cup ParRoles$ be the set of roles.

Finally, we let $O = \mathit{Users} \cup \mathit{Items} \cup G \cup R$ be the set of objects. Access to objects is regulated by permissions, drawn from a set Perms. The *scope* of a granted permission can be narrowed or extended using a flag $s \in \{-, \downarrow\}$, which specifies if a permission is given over an individual object or if it can be inherited through the group hierarchy.

Definition 1 (System). *A system is a tuple* $\mathcal{S} = (PR, GR, U, I, UR, UG, UU)$:

- $PR \subseteq (\mathit{RegRoles} \times \mathit{Perms} \times O \times \{-, \downarrow\}) \cup (\mathit{RoleTemps} \times \mathit{Perms})$ *is the permission-assignment relation;*
- $GR \subseteq G \times \mathit{RegRoles}$ *is a relation mapping groups to regular roles;*
- $U \subseteq \mathit{Users}$ *is a finite set of users;*
- $I \subseteq \mathit{Items}$ *is a finite set of items;*
- $UR \subseteq U \times R$ *is a relation mapping users to their assigned roles;*
- $UG \subseteq U \times G$ *is a relation mapping users to the groups they belong to;*
- $UU \subseteq U \times U$ *is a relation mapping users to users, modelling impersonation.*

By convention we assume that $\forall p \in \mathit{Perms} : (\mathsf{Owner}_0[\cdot], p) \in PR$, *i.e., the first owner role template in the system has full permissions.*

2.2 Semantics

As per previous studies [3,9], we find it convenient to decouple a system $\mathcal{S}$ into a static *policy* $\mathcal{P}$ and a dynamic *configuration* σ. A policy is a pair $\mathcal{P} = (PR, GR)$, while a configuration is a 5-tuple $\sigma = (U, I, UR, UG, UU)$. The reduction semantics of Liferay RBAC has then the form $\mathcal{P} \vdash \sigma \xrightarrow{\beta} \sigma'$ for some label β.

To specify the semantics, we start by defining to which groups is assigned a given object o under a user-to-group mapping UG. Formally, we inductively define the set $\mathit{groups}_{UG}(o)$ through the self-explanatory inference rules in Table 1.

Table 1. Group Assignment

(G-ITEM)	(G-USER)	(G-GROUP)	(G-INHERIT)
$g \in i.\mathit{groups}$	$(u, g) \in UG$		$g' \preceq g$ $g' \in \mathit{groups}_{UG}(o)$
$g \in \mathit{groups}_{UG}(i)$	$g \in \mathit{groups}_{UG}(u)$	$g \in \mathit{groups}_{UG}(g)$	$g \in \mathit{groups}_{UG}(o)$

We then define when a user u is granted a permission p over an object o in the system $\mathcal{S}$. The definition of the judgement $\mathcal{S} \vdash \mathit{granted}(u, p, o)$ is in Table 2.

Rule (P-REGI) is standard: it states that, if a user u has a regular role r which grants permission p on the object o, then u has p on o. Rule (P-REGG) allows to extend a permission given over a group g to any object o belonging to g: notice that this must be made explicit in the policy, by using the flag $\downarrow$ when assigning the permission. Rules (P-GROUPI) and (P-GROUPG) are

Table 2. Permission Granting, where $\mathcal{S} = (PR, GR, U, I, UR, UG, UU)$

(P-REGI)	(P-REGG)	(P-GROUPI)
$(u, r) \in UR$	$(u, r) \in UR$	$(u, g) \in UG$
$(r, p, o, -) \in PR$	$(r, p, g, \downarrow) \in PR$	$(g, r) \in GR$
	$g \in \mathit{groups}_{UG}(o)$	$(r, p, o, -) \in PR$
$\mathcal{S} \vdash \mathit{granted}(u, p, o)$	$\mathcal{S} \vdash \mathit{granted}(u, p, o)$	$\mathcal{S} \vdash \mathit{granted}(u, p, o)$

(P-GROUPG)

$$\frac{(u, g) \in UG \quad (g, r) \in GR \quad (r, p, g', \downarrow) \in PR \quad g' \in \mathit{groups}_{UG}(o)}{\mathcal{S} \vdash \mathit{granted}(u, p, o)}$$

(P-TEMPLATE)

$$\frac{(u, r[g]) \in UR \quad (r[\cdot], p) \in PR \quad g \in \mathit{groups}_{UG}(u) \cap \mathit{groups}_{UG}(o)}{\mathcal{S} \vdash \mathit{granted}(u, p, o)}$$

(P-OWNER)

$$\frac{(u, \mathsf{Owner}_j[i]) \in UR \quad (\mathsf{Owner}_j[\cdot], p) \in PR}{\mathcal{S} \vdash \mathit{granted}(u, p, i)}$$

(P-IMPERSONATE)

$$\frac{(u, u') \in UU \quad \mathcal{S}[UU \mapsto \emptyset] \vdash \mathit{granted}(u', p, o)}{\mathcal{S} \vdash \mathit{granted}(u, p, o)}$$

Convention: for any u and i occurring in the judgement we require $u \in U$ and $i \in I$

the counterparts of (P-REGI) and (P-REGG) for (regular) roles assigned to groups: they state that any role given to a group is inherited by any user in that group.

Rule (P-TEMPLATE) is subtle and defines the semantics of parametrized roles: if a role template $r[\cdot]$ is given the permission p and the parametrized role $r[g]$ is assigned to a given user u, then u has p on any object in g, provided that u is himself a member of that group. In this way, a parametrized role $r[g]$ allows to constrain the scope of a permission p inside the group g.

Rule (P-OWNER) formalizes the intuition behind owner roles: if a role template $\mathsf{Owner}_j[\cdot]$ is given the permission p and the parametrized role $\mathsf{Owner}_j[i]$ is assigned to a user u for some item i, then u has p on i. Notice that, by the convention in Definition 1, a user with role $\mathsf{Owner}_0[i]$ has full permissions on i.

Finally, rule (P-IMPERSONATE) deals with permissions which are granted upon impersonation: if u is impersonating u' and u' has permission p on the object o, then u has p on o. There is a subtle point to notice though: if u is impersonating u' and u' is impersonating u'', then u is *not* granted the permissions of u''. Formally, this is ensured by emptying the UU component of $\mathcal{S}$ before deriving the judgement in the premises of rule (P-IMPERSONATE). In Liferay, only the permissions which are *statically* known to be granted to u' are inherited by a user impersonating u'.

Having defined when a user is granted a given permission, the formal semantics is relatively simple. The reduction rules are given in Table 4, we just comment the most interesting points. First, owner roles can only be assigned when new items are created and are only removed when items are deleted; we conservatively assume that the owners of dynamically created items have full

Table 3. Reduction Semantics of Liferay RBAC

(ASSIGN-ROLE)

$$\frac{\mathcal{S} \vdash granted(u_1, \mathsf{AssignRole}, r) \qquad r \neq \mathsf{Owner}_j[i]}{\mathcal{P} \vdash (U, I, UR, UG, UU) \xrightarrow{assign\_role(u_1, u_2, r)} (U, I, UR \cup \{(u_2, r)\}, UG, UU)}$$

(REMOVE-ROLE)

$$\frac{\mathcal{S} \vdash granted(u_1, \mathsf{RemoveRole}, r) \qquad r \neq \mathsf{Owner}_j[i]}{\mathcal{P} \vdash (U, I, UR, UG, UU) \xrightarrow{remove\_role(u_1, u_2, r)} (U, I, UR \setminus \{(u_2, r)\}, UG, UU)}$$

(ASSIGN-GROUP)

$$\frac{\mathcal{S} \vdash granted(u_1, \mathsf{AssignGroup}, g)}{\mathcal{P} \vdash (U, I, UR, UG, UU) \xrightarrow{assign\_group(u_1, u_2, g)} (U, I, UR, UG \cup \{(u_2, g)\}, UU)}$$

(REMOVE-GROUP)

$$\frac{\mathcal{S} \vdash granted(u_1, \mathsf{RemoveGroup}, g)}{\mathcal{P} \vdash (U, I, UR, UG, UU) \xrightarrow{remove\_group(u_1, u_2, g)} (U, I, UR, UG \setminus \{(u_2, g)\}, UU)}$$

(IMPERSONATE)

$$\frac{u_1 \notin dom(UU) \qquad \mathcal{S} \vdash granted(u_1, \mathsf{Impersonate}, u_2)}{\mathcal{P} \vdash (U, I, UR, UG, UU) \xrightarrow{impersonate(u_1, u_2)} (U, I, UR, UG, UU \cup \{(u_1, u_2)\})}$$

(DEIMPERSONATE)

$$\mathcal{P} \vdash (U, I, UR, UG, UU) \xrightarrow{deimpersonate(u_1, u_2)} (U, I, UR, UG, UU \setminus \{(u_1, u_2)\})$$

(ADD-USER)

$$\mathcal{P} \vdash (U, I, UR, UG, UU) \xrightarrow{add\_user(u)} (U \cup \{u\}, I, UR, UG, UU)$$

(REMOVE-USER)

$$\frac{u \notin dom(UU) \qquad UR' = \{(u', r) \in UR \mid u' \neq u\} \qquad UG' = \{(u', g) \in UG \mid u' \neq u\}}{\mathcal{P} \vdash (U, I, UR, UG, UU) \xrightarrow{remove\_user(u)} (U \setminus \{u\}, I, UR', UG', UU)}$$

(ADD-ITEM)

$$\frac{\mathcal{S} \vdash granted(u, \mathsf{AddItem}, g) \qquad g \in i.groups}{\mathcal{P} \vdash (U, I, UR, UG, UU) \xrightarrow{add\_item(u, i)} (U, I \cup \{i\}, UR \cup \{(u, \mathsf{Owner}_0[i])\}, UG, UU)}$$

(REMOVE-ITEM)

$$\frac{\mathcal{S} \vdash granted(u, \mathsf{RemoveItem}, i) \qquad UR' = \{(u', r) \in UR \mid r \neq \mathsf{Owner}_j[i]\}}{\mathcal{P} \vdash (U, I, UR, UG, UU) \xrightarrow{remove\_item(u, i)} (U, I \setminus \{i\}, UR', UG, UU)}$$

Notation: we assume $\mathcal{P} = (PR, GR)$ and $\mathcal{S} = (PR, GR, U, I, UR, UG, UU)$

permissions on them, by using the role template $\mathsf{Owner}_0[\cdot]$ in rule (ADD-ITEM). We then notice that each user can only impersonate a single user at a time, by the side-condition $u_1 \notin dom(UU)$ in rule (IMPERSONATE); this implicitly ensures that impersonation is *not* transitive, i.e., if u is impersonating u' and u' can impersonate u'', then u cannot impersonate u''. Finally, when removing a user u, we require that u is not impersonating anyone: this is technically convenient and not limiting, since we can always apply rule (DEIMPERSONATE) up to a configuration where u is impersonating none and then remove him. Notice that no permission is needed to deimpersonate an impersonated user.

We write $\mathcal{P} \vdash \sigma \xrightarrow{\vec{\beta}} \sigma'$ if and only if there exist $\sigma_1, \ldots, \sigma_{n-1}$ such that $\mathcal{P} \vdash \sigma \xrightarrow{\beta_1} \sigma_1 \wedge \mathcal{P} \vdash \sigma_1 \xrightarrow{\beta_2} \sigma_2 \wedge \ldots \wedge \mathcal{P} \vdash \sigma_{n-1} \xrightarrow{\beta_n} \sigma'$ for some $\vec{\beta} = \beta_1, \ldots, \beta_n$.

3   Verification of Liferay RBAC

The formal semantics in the previous section can be useful to spot improper privilege escalations by untrusted users of the portal, but it cannot be directly used to prove the *absence* of undesired accesses by an exhaustive state space exploration, since the corresponding labelled transition system has an infinite number of states. We now discuss how we tackle the problem of policy verification by *abstract model-checking* [6].

3.1   A Modal Logic for Verification

We let the syntax of formulas be defined by the following productions:

$$\begin{aligned} \textit{State formulas} \quad & \phi ::= granted(u, p, o) \mid \phi \wedge \phi \mid \phi \vee \phi, \\ \textit{Path formulas} \quad & \varphi ::= \Diamond \phi \mid \neg \varphi \mid \varphi \wedge \varphi \mid \varphi \vee \varphi. \end{aligned}$$

This is a simple modal logic, where the modality $\Diamond$ is equivalent to the "finally" operator F available in full-fledged temporal logics like CTL, CTL* or LTL [5]. The (standard) satisfaction relations for state formulas and path formulas are defined by the judgements $\mathcal{P}, \sigma \models \phi$ and $\mathcal{P}, \sigma \models \varphi$ in Table 4. The path formula $\Diamond \phi$ is satisfied by $\mathcal{P}, \sigma$ whenever there exists a reachable configuration from σ under $\mathcal{P}$ where the state formula ϕ holds true.

Though simple, the logic above allows to formalize several standard security properties of interest for RBAC systems. For instance, we have:

– role reachability: user u can never be assigned to regular role r:

$$\mathcal{P}, \sigma \models \neg \Diamond (granted(u, p^*, o^*)),$$

where p^* is a dummy permission on a dummy object o^* assigned only to r;
– mutual exclusion: user u can never possess both p and p' on object o:

$$\mathcal{P}, \sigma \models \neg \Diamond (granted(u, p, o) \wedge granted(u, p', o))$$

Table 4. Satisfaction Relation

$$\frac{\text{(LS-BASIC)}}{\mathcal{P}, \sigma \vdash granted(u, p, o)}{\mathcal{P}, \sigma \models granted(u, p, o)} \qquad \frac{\text{(LS-AND)}}{\mathcal{P}, \sigma \models \phi_1 \quad \mathcal{P}, \sigma \models \phi_2}{\mathcal{P}, \sigma \models \phi_1 \wedge \phi_2} \qquad \frac{\text{(LS-OR)}}{\mathcal{P}, \sigma \models \phi_i}{\mathcal{P}, \sigma \models \phi_1 \vee \phi_2}$$

$$\frac{\text{(LP-FINALLY)}}{\mathcal{P} \vdash \sigma \overset{\vec{s}}{\Rightarrow} \sigma' \quad \mathcal{P}, \sigma' \models \phi}{\mathcal{P}, \sigma \models \Diamond \phi} \qquad \frac{\text{(LP-NOT)}}{\mathcal{P}, \sigma \not\models \varphi}{\mathcal{P}, \sigma \models \neg \varphi} \qquad \frac{\text{(LP-AND)}}{\mathcal{P}, \sigma \models \varphi_1 \quad \mathcal{P}, \sigma \models \varphi_2}{\mathcal{P}, \sigma \models \varphi_1 \wedge \varphi_2} \qquad \frac{\text{(LP-OR)}}{\mathcal{P}, \sigma \models \varphi_i}{\mathcal{P}, \sigma \models \varphi_1 \vee \varphi_2}$$

Notation: in rules (LS-OR) and (LP-OR) we let $i \in \{1, 2\}$

– group reachability: user u can never join group g:

$$\mathcal{P}, \sigma \models \neg \Diamond (granted(u, p^*, g)),$$

where p^* is a dummy permission assigned only to the dummy role template $r[\cdot]$ and the parametrized role $r[g]$ is assigned to u in the configuration σ.

3.2   Abstract Semantics

The abstract semantics builds on two core ideas. First, we observe that the actual identity of users and items is often immaterial: for instance, two items belongings to the same groups behave exactly in the same way for Liferay RBAC. Second, many transitions of the semantics (e.g., removing roles) actually weaken the privileges granted to a given user, hence they are irrelevant to detect security violations. Leveraging these two observations, the abstract semantics consists of: (i) a finite-range abstraction function $\alpha : O \to O$, mapping each object in the unbounded set O to some canonical representative; and (ii) an abstract reduction relation, defining the dynamics of configurations abstracted by α. The abstraction function can be chosen arbitrarily, as long as it satisfies some syntactic conditions given below: one may choose different trade-offs between precision and efficiency by using different abstractions for verification.

We presuppose two functions $\alpha_u : Users \to Users$ and $\alpha_i : Items \to Items$. We then build on top of them an abstraction function $\alpha : O \to O$ as follows:

$$\alpha(o) = \begin{cases} \alpha_u(u) & \text{if } o \text{ is a user } u \\ \alpha_i(i) & \text{if } o \text{ is an item } i \\ r[\alpha_i(i)] & \text{if } o \text{ is a parametrized role } r[i] \\ o & \text{otherwise.} \end{cases}$$

We extend α to formulas, (sequences of) labels, tuples and sets by applying it to any object syntactically occurring therein.

The *abstract* reduction relation $\mathcal{P} \vdash \sigma \overset{\alpha}{\mapsto} \sigma'$ is obtained from the rules in Table 4 by excluding (REMOVE-ROLE), (REMOVE-GROUP), (REMOVE-USER) and (REMOVE-ITEM), and by dropping the side-condition $u_1 \notin dom(UU)$ from rule (IMPERSONATE).

We let $\mathcal{P} \vdash \sigma \overset{\vec{\beta}}{\Longmapsto} \sigma'$ be the obvious generalization to the abstract semantics of the relation $\mathcal{P} \vdash \sigma \overset{\vec{\beta}}{\Rightarrow} \sigma'$ defined above. We then let $\mathcal{P}, \sigma \models_\alpha \varphi$ be the satisfaction relation obtained from the rules in Table 4 by replacing (LP-FINALLY) with:

(ALP-FINALLY)

$$\frac{\alpha(\mathcal{P}) \vdash \alpha(\sigma) \overset{\alpha(\vec{\beta})}{\Longmapsto} \sigma' \qquad \alpha(\mathcal{P}), \sigma' \models \alpha(\phi)}{\mathcal{P}, \sigma \models_\alpha \Diamond \phi}$$

and by introducing the obvious counterparts of rules (LP-NOT), (LP-AND) and (LP-OR). Since any abstraction function has a finite range, it is easy to prove:

Lemma 1. *There exists a decision procedure* ABS-SAT$(\alpha, \mathcal{P}, \sigma, \varphi)$ *for* $\mathcal{P}, \sigma \models_\alpha \varphi$.

We verify security properties of the infinite-state concrete semantics by model-checking the finite-state abstract semantics. To isolate the fragment of the modal logic amenable for verification, we let negation-free formulas $\widehat{\varphi}$ and rank-1 formulas ψ be defined by the following productions:

$$\begin{aligned} \text{Negation-free formulas} \quad & \widehat{\varphi} ::= \Diamond \phi \mid \widehat{\varphi} \wedge \widehat{\varphi} \mid \widehat{\varphi} \vee \widehat{\varphi}, \\ \text{Rank-1 formulas} \quad & \psi ::= \neg \widehat{\varphi} \mid \psi \wedge \psi \mid \psi \vee \psi. \end{aligned}$$

We can construct a procedure which determines if an arbitrary rank-1 formula is satisfied or not by the concrete semantics (see Fig. 1). The next subsections discuss sufficient conditions for the soundness and the completeness of the algorithm, i.e., conditions on the policy $\mathcal{P}$ and the abstraction function α which ensure that a positive (resp. negative) answer by SAT$(\alpha, \mathcal{P}, \sigma, \psi)$ implies that ψ is satisfied (resp. not satisfied) by $\mathcal{P}, \sigma$. Notice that all the example properties previously described are expressed by a rank-1 formula.

```
SAT(α, P, σ, ψ):
    match ψ with
    | ¬φ̂        →  not ABS-SAT(α, P, σ, φ̂)
    | ψ₁ ∧ ψ₂   →  SAT(α, P, σ, ψ₁) and SAT(α, P, σ, ψ₂)
    | ψ₁ ∨ ψ₂   →  SAT(α, P, σ, ψ₁) or SAT(α, P, σ, ψ₂)
```

Fig. 1. Abstract Model-Checking Algorithm

3.3   Soundness of Verification

We first prove the *soundness* of the algorithm in Fig. 1, i.e., we show that, assuming a mild syntactic restriction on the abstraction function α, a positive answer by $\text{SAT}(\alpha, \mathcal{P}, \sigma, \psi)$ implies that $\mathcal{P}, \sigma \models \psi$ holds true.

Definition 2 (Group-preserving Abstraction). *An abstraction function α is* group-preserving *iff $\forall i \in Items : i.groups \subseteq \alpha(i).groups$.*

Definition 3 (Permission-based Ordering). *We let $\sigma \sqsubseteq_\mathcal{P} \sigma'$ if and only if $\mathcal{P}, \sigma \vdash granted(u, p, o)$ implies $\mathcal{P}, \sigma' \vdash granted(u, p, o)$ for any u, p and o.*

The next theorem states that any behaviour of the concrete semantics has a counterpart in the abstract semantics. It also ensures that the abstract semantics over-approximates the permissions granted to each user. The result would not hold in general if the side-condition of rule (IMPERSONATE) was included in the abstract semantics: we omit further technical details due to space constraints.

Theorem 1 (Soundness). *Let α be group-preserving. If $\mathcal{P} \vdash \sigma \xrightarrow{\vec{\beta}} \sigma'$, then there exists a sub-trace of $\vec{\beta}$, call it $\vec{\gamma}$, such that $\alpha(\mathcal{P}) \vdash \alpha(\sigma) \xrightarrow{\alpha(\vec{\gamma})} \sigma''$ for some σ'' such that $\alpha(\sigma') \sqsubseteq_{\alpha(\mathcal{P})} \sigma''$.*

Using the theorem above, we can prove the soundness of verification. Notice that the only assumption needed for soundness is on the abstraction function α: the result applies to any choice of $\mathcal{P}$, σ and ψ.

Theorem 2 (Sound Verification). *Let α be a group-preserving abstraction function. If $\text{SAT}(\alpha, \mathcal{P}, \sigma, \psi)$ returns a positive answer, then $\mathcal{P}, \sigma \models \psi$.*

3.4   Completeness of Verification

We now identify conditions for the *completeness* of the algorithm in Fig. 1, i.e., we discuss under which assumptions a negative answer by $\text{SAT}(\alpha, \mathcal{P}, \sigma, \psi)$ ensures that $\mathcal{P}, \sigma \not\models \psi$. This is important for the *precision* of the analysis.

To state and prove the completeness result, we focus on a particular class of policies which does not allow to impersonate users. We remark that this corresponds to a realistic use case, since Liferay can be configured to prevent impersonation by setting the property `portal.impersonation.enable` to false in the file `webapps/ROOT/WEB-INF/classes/portal-developer.properties`.

Definition 4 (Impersonation-free Policy/Configuration). *A policy $\mathcal{P} = (PR, GR)$ is* impersonation-free *iff $\forall(r, p, o, s) \in PR : p \neq$ Impersonate and $\forall(r[\cdot], p) \in PR : p \neq$ Impersonate. A configuration $\sigma = (U, I, UR, UG, UU)$ is* impersonation-free *iff $UU = \emptyset$.*

Given a configuration $\sigma = (U, I, UR, UG, UU)$, let: $users(\sigma) = U$; $items(\sigma) = I$; $groups_\sigma(u) = \{g \mid (u, g) \in UG\}$; and $roles_\sigma(u) = \{r \mid (u, r) \in UR\}$.

The completeness result requires to find sufficient conditions which ensure that the abstract semantics is under-approximating the concrete semantics. If impersonation is never used, it is enough to require that each user in the abstract semantics belongs to fewer groups and has fewer roles than one of his corresponding users in the concrete semantics, as formalized next.

Definition 5 (Abstract Under-Approximation). *A configuration σ is an abstract under-approximation of a configuration σ', written $\sigma \lesssim_\alpha \sigma'$, if and only if both the following conditions hold true:*

- $\forall u \in users(\sigma) : \exists u' \in users(\sigma') : u = \alpha(u') \wedge groups_\sigma(u) \subseteq groups_{\sigma'}(u') \wedge roles_\sigma(u) \subseteq \alpha(roles_{\sigma'}(u'))$;
- $\forall i \in items(\sigma) : \exists i' \in items(\sigma') : i = \alpha(i')$.

Definition 6 (Group-forgetting Abstraction). *An abstraction function α is group-forgetting iff $\forall i \in Items : \alpha(i).groups \subseteq i.groups$.*

The next theorem states that any behaviour in the abstract semantics has a counterpart in the concrete semantics, assuming that impersonation is never used. It additionally ensures that the desired under-approximation is preserved upon reduction.

Theorem 3 (Completeness). *Let $\mathcal{P}, \sigma$ be impersonation-free and let α be group-forgetting. If $\alpha(\sigma) \lesssim_\alpha \sigma$ and $\alpha(\mathcal{P}) \vdash \alpha(\sigma) \xmapsto{\alpha(\vec{\beta})} \sigma'$ for some $\vec{\beta}$ and σ', then there exists σ'' such that $\mathcal{P} \vdash \sigma \xRightarrow{\vec{\beta}} \sigma''$ and $\sigma' \lesssim_\alpha \sigma''$.*

We need also an additional condition to prove the completeness of verification: we must ensure that the identity of any object occurring in the formula ψ to verify is *respected* by the abstraction function, in the following sense.

Definition 7 (Respectful Abstraction). *An abstraction function α respects an object o iff $\alpha(o) = o$ and $\forall o' \in O : o' \neq o \Rightarrow \alpha(o') \neq o$. An abstraction function α respects ψ iff it respects any object occurring in ψ.*

Theorem 4 (Complete Verification). *Let $\mathcal{P}, \sigma$ be impersonation-free and let α be a group-forgetting abstraction function which respects ψ. If $\alpha(\sigma) \lesssim_\alpha \sigma$ and $\mathcal{P}, \sigma \models \psi$, then $\mathrm{SAT}(\alpha, \mathcal{P}, \sigma, \psi)$ returns a positive answer.*

Completeness of verification does not hold in general if impersonation is used. Specifically, if a user u is allowed to impersonate both u_1 and u_2, he will be able to do it at the same time in the abstract semantics, thus getting the union of their privileges; however, the two users cannot be impersonated at the same time in the concrete semantics, which breaks the intended under-approximation.

4   Implementation: LifeRBAC

LIFERBAC is a Liferay plugin providing a simple user interface to let portal administrators input security queries about the underlying RBAC system. The plugin takes a snapshot of the portal and translates it into a corresponding representation in our formal model, encoded in the ASLAN specification language [18]. The initial state representation is joined with a set of (hand-coded) transition rules, corresponding to the ASLAN implementation of the abstract semantics, and the query is translated into a modal logic formula, which is verified using the state-of-art model-checker SATMC [1].

4.1   Implementation Details

LIFERBAC currently supports two different analyses, corresponding to the choice of two different abstraction functions. In the *fast* analysis, only the identity of the users occurring in the security query is preserved by the abstraction function, while all the other users are abstracted into a super-user with the union of their privileges. In the *precise* analysis, the identity of the users occurring in the security query is still preserved, but all the other users are abstracted into a canonical representative sharing their same groups and roles. As to items, in both cases we preserve their identity when they occur in the security query, while we abstract them into a canonical representative sharing the same groups otherwise. Observe that both the choices of the abstraction function satisfy the conditions of Theorem 2, hence both the analyses are *sound*. Moreover, the precise analysis satisfies also the conditions of Theorem 4, hence it is *complete* for any policy which does not allow to impersonate users.

At the moment LIFERBAC only supports the verification of security queries predicating over a subset of the objects available in Liferay, i.e., users, groups and layouts. Including additional types of objects (e.g., portlets) is essentially a matter of programming.

4.2   Experiments

Inspired by a previously published case study [23], we consider an experimental setting modelling a hypothetical university with 3 departments and 10 courses. We represent the university as the only company in the portal and the departments as three different organizations; then, we create a private site for each department and a corresponding child site for every course. We consider 15 role templates: 6 templates are used to generate site roles, while 9 templates are used to generate organization roles. We also include 14 regular roles to collect permissions which are not scoped to any specific site or organization. Overall, we have $6 \cdot 13 + 9 \cdot 3 + 14 = 119$ roles; for each of them, we model access permissions to different resources as read/write privileges on specific web pages in the sites, and we enable administrative permissions where appropriate, e.g., a user with role Professor[c] for some course c can assign the parametrized role Student[c]. Finally, we create 1000 users with different role combinations.

In the experiments, we consider three different security queries:

- q_1: can student u_1, who is a member of site s_1, delete a page from s_1?
- q_2: can student u_2 join site s_2 belonging to another department?
- q_3: can student u_3, who is a member of site s_3, assign a user u_4 to s_3?

All the three queries are performed against three slightly different configurations of increasing complexity. In the first configuration no user is allowed to impersonate other users or to assign members to existing sites/organizations. In the second configuration clerks can move users along the group hierarchy, but no impersonation is possible, while in the third configuration administrators are allowed to impersonate any user. The last scenario is a "stress test" for our tool, since unconstrained impersonation leads to a state-space explosion.

The time required to check the three queries against the three described configurations and their results are given in Table 5. In the table, we also keep track of the number of users obtained after applying the abstraction; all the attacks have been confirmed by hand on Liferay 6.2 CE. The experiments have been performed on an Intel Xeon 2.4Ghz running Ubuntu Linux 12.04 LTS.

Table 5. Experimental Results

conf.	query	Fast Analysis				Precise Analysis			
		#users	time	attack	real	#users	time	attack	real
A	q_1	3	1m 58s	yes	yes	62	6m 10s	yes	yes
	q_2	3	1m 46s	no	-	62	1m 52s	no	-
	q_3	3	1m 48s	no	-	62	1m 53s	no	-
B	q_1	3	2m 40s	yes	yes	62	49m 01s	yes	yes
	q_2	3	1m 50s	yes	yes	62	21m 46s	yes	yes
	q_3	3	1m 50s	no	-	62	22m 30s	no	-
C	q_1	3	2m 53s	yes	yes	62	182m 21s	yes	yes
	q_2	3	1m 59s	yes	yes	62	56m 02s	yes	yes
	q_3	3	2m 37s	no	-	62	54m 46s	no	-

A = no impersonate and no groups; B = only groups; C = groups and impersonate
#users = the number of users after the abstraction; real = attack confirmed

For the first configuration, where we do not include Liferay-specific features, the verification time is very good and in line with previous work on standard RBAC systems [9]. In the worst case, verification takes around 6 minutes and we are able to *prove* the existence of an attack by the completeness result, something which is beyond previous abstraction-based proposals. Based on the numbers in the table, we observe that the possibility of assigning and removing groups, thus activating and deactivating parametrized roles, has a significant impact on the performances, especially for the precise analysis. Impersonation further contributes to complicate the verification problem, as the results for the precise analysis clearly highlight, but it does not hinder too much the performances of the fast analysis. Remarkably, despite the huge approximation applied by the fast analysis, we did not identify false positives for these queries (and also for other tests we performed). We leave as future work further study on the precise analysis, to improve its performance by the usage of static slicing techniques [8].

4.3   Discussion: Adequacy of the Semantics

A thorny issue when verifying a complex framework like Liferay RBAC is bridging the gap between the formal model and the real system. In particular, observe that: (*i*) Liferay features many different permissions and it is not obvious which ones correspond to the few administrative permissions (e.g., AssignRole) we include in the model; and (*ii*) the Liferay permission checker is a complicated piece of code, which selectively grants or denies some permissions based on the *type* of the object of an access request, but types have only a marginal role in the permission granting process formalized in Table 2.

The key insight is that both the problems above can be dealt with just by carefully constructing the permission-assignment relation *PR* used in the formal model. We assessed the adequacy of our solution by *testing*, an idea proposed by the programming languages community [12]. Specifically, we created a tool, LR-TEST, which takes a snapshot of the Liferay portal and constructs its corresponding representation in our formal model. The tool systematically queries both the Liferay permission checker and its formal counterpart (deriving the judgements in Table 2) to detect any mismatch on resource accesses and enabled administrative actions. Mismatches may be of two types: false positives lead to an over-conservative security analysis, while false negatives lead to overlooking real security flaws. In the current prototype, we eliminated all the false negatives we identified and we only left a few false positives to be fixed.

5   Related Work

Abstraction techniques as an effective tool for RBAC verification have been first concurrently proposed by Bugliesi *et al.* [3] and Ferrara *et al.* [9]. The first paper proposes a formal semantics for grsecurity, an RBAC system built on top of the Linux kernel, and then uses abstract model-checking to verify some specific security properties of interest. Notably, the abstraction adopted in [3] is fixed, while the results in this paper are parametric with respect to the choice of an abstraction function. The authors of [9], instead, do not apply model-checking techniques, but rather construct an imperative program abstracting the dynamics of the RBAC system and apply standard results from program verification to soundly approximate the role reachability problem. The proposed abstraction is incomplete, i.e., the analysis may produce false positives. In recent work [8], the same research group presented VAC, a tool for role reachability analysis, which can be used both to prove RBAC policies correct and to find errors in them (adopting different analysis backends).

A different approach to RBAC verification is based on SMT solving. Armando and Ranise proposed a symbolic verification framework for RBAC systems with an unbounded number of users [2]. More recent research holds great promise in making similar techniques scale to verify large RBAC policies [19].

Error finding in RBAC policies is complementary to verification and abstraction techniques have proved fundamental also for this problem, since plain model-checking does not scale to the analysis of real systems. The most known proposal in the area is MOHAWK, a tool based on an abstraction-refinement model-checking strategy [13]. Interestingly, MOHAWK has been recently extended to prove RBAC policies correct [14]. The analysis is limited to separate administration policies, a restriction which we do not assume in this paper.

To overcome the performance issues affecting RBAC policy verification, many authors identified tractable fragments of the general RBAC model and presented different algorithms to answer useful security queries under specific policy restrictions [23,11,21,16]. Most of these results are proved for a finite set of users, while our work assumes an unbounded set of users entering and leaving the system.

6 Conclusion

We presented a formal semantics for Liferay RBAC and we tackled the verification problem through abstract model-checking, discussing sufficient conditions for the soundness and the completeness of the analysis. We then implemented a tool, LIFERBAC, which can be used to verify the security of real Liferay portals and we reported on experiments showing the effectiveness of our solution.

As a future work, we plan to strengthen the completeness result to policies involving impersonation. Moreover, we want to extend LIFERBAC to support error finding in the underlying RBAC policy, to include a counter-example generation module for flawed policies, and to make it significantly faster by adapting static slicing techniques from the literature [8].

Acknowledgements. This research is partially supported by the MIUR projects ADAPT and CINA. Alvise Rabitti acknowledges support by SMC Treviso. We would like to thank Rocco Germinario and Paolo Valeri of SMC Treviso for participating to many technical discussions about Liferay.

References

1. Armando, A., Carbone, R., Compagna, L.: SATMC: A SAT-based model checker for security-critical systems. In: Ábrahám, E., Havelund, K. (eds.) TACAS 2014 (ETAPS). LNCS, vol. 8413, pp. 31–45. Springer, Heidelberg (2014)
2. Armando, A., Ranise, S.: Automated symbolic analysis of ARBAC-policies. In: Cuellar, J., Lopez, J., Barthe, G., Pretschner, A. (eds.) STM 2010. LNCS, vol. 6710, pp. 17–34. Springer, Heidelberg (2011)
3. Bugliesi, M., Calzavara, S., Focardi, R., Squarcina, M.: Gran: Model checking gr-security RBAC policies. In: Computer Security Foundations (CSF), pp. 126–138 (2012)
4. Calzavara, S., Rabitti, A., Bugliesi, M.: Formal verification of Liferay RBAC (full version), www.dais.unive.it/~calzavara/papers/essos15-full.pdf

5. Clarke, E.M., Emerson, E.A., Sistla, A.P.: Automatic verification of finite-state concurrent systems using temporal logic specifications. ACM Trans. Program. Lang. Syst. 8(2), 244–263 (1986)
6. Cousot, P., Cousot, R.: Refining model checking by abstract interpretation. Autom. Softw. Eng. 6(1), 69–95 (1999)
7. Ferraiolo, D.F., Sandhu, R.S., Gavrila, S.I., Kuhn, D.R., Chandramouli, R.: Proposed NIST standard for role-based access control. ACM Trans. Inf. Syst. Secur. 4(3), 224–274 (2001)
8. Ferrara, A.L., Madhusudan, P., Nguyen, T.L., Parlato, G.: VAC - verifier of administrative role-based access control policies. In: Biere, A., Bloem, R. (eds.) CAV 2014. LNCS, vol. 8559, pp. 184–191. Springer, Heidelberg (2014)
9. Ferrara, A.L., Madhusudan, P., Parlato, G.: Security analysis of role-based access control through program verification. In: Computer Security Foundations (CSF), pp. 113–125 (2012)
10. Giuri, L., Iglio, P.: Role templates for content-based access control. In: ACM Workshop on Role-Based Access Control, pp. 153–159 (1997)
11. Gofman, M.I., Luo, R., Solomon, A.C., Zhang, Y., Yang, P., Stoller, S.D.: RBAC-PAT: A policy analysis tool for role based access control. In: Kowalewski, S., Philippou, A. (eds.) TACAS 2009. LNCS, vol. 5505, pp. 46–49. Springer, Heidelberg (2009)
12. Guha, A., Saftoiu, C., Krishnamurthi, S.: The essence of JavaScript. In: D'Hondt, T. (ed.) ECOOP 2010. LNCS, vol. 6183, pp. 126–150. Springer, Heidelberg (2010)
13. Jayaraman, K., Ganesh, V., Tripunitara, M.V., Rinard, M.C., Chapin, S.J.: Automatic error finding in access-control policies. In: ACM Conference on Computer and Communications Security (CCS), pp. 163–174 (2011)
14. Jayaraman, K., Tripunitara, M.V., Ganesh, V., Rinard, M.C., Chapin, S.J.: Mohawk: Abstraction-refinement and bound-estimation for verifying access control policies. ACM Trans. Inf. Syst. Secur. 15(4), 18 (2013)
15. Li, N., Mitchell, J.C.: A role-based trust-management framework. In: DARPA Information Survivability Conference and Exposition (DISCEX), pp. 201–212 (2003)
16. Li, N., Tripunitara, M.V.: Security analysis in role-based access control. ACM Trans. Inf. Syst. Secur. 9(4), 391–420 (2006)
17. Liferay Inc.: Liferay clients and case studies,
    https://www.liferay.com/it/products/liferay-portal/stories
18. Mödersheim, S.: Deciding security for a fragment of ASLan. In: Foresti, S., Yung, M., Martinelli, F. (eds.) ESORICS 2012. LNCS, vol. 7459, pp. 127–144. Springer, Heidelberg (2012)
19. Ranise, S., Truong, A., Armando, A.: Boosting model checking to analyse large ARBAC policies. In: Jøsang, A., Samarati, P., Petrocchi, M. (eds.) STM 2012. LNCS, vol. 7783, pp. 273–288. Springer, Heidelberg (2013)
20. Sandhu, R.S., Bhamidipati, V., Munawer, Q.: The ARBAC97 model for role-based administration of roles. ACM Trans. Inf. Syst. Secur. 2(1), 105–135 (1999)
21. Sasturkar, A., Yang, P., Stoller, S.D., Ramakrishnan, C.R.: Policy analysis for administrative role-based access control. Theor. Comput. Sci. 412(44), 6208–6234 (2011)
22. Stoller, S.D., Yang, P., Gofman, M.I., Ramakrishnan, C.R.: Symbolic reachability analysis for parameterized administrative role-based access control. Computers & Security 30(2-3), 148–164 (2011)
23. Stoller, S.D., Yang, P., Ramakrishnan, C.R., Gofman, M.I.: Efficient policy analysis for administrative role based access control. In: ACM Conference on Computer and Communications Security (CCS), pp. 445–455 (2007)

Formal Verification of Privacy Properties in Electric Vehicle Charging

Marouane Fazouane[1], Henning Kopp[2], Rens W. van der Heijden[2],
Daniel Le Métayer[1], and Frank Kargl[2]

[1] Inria, University of Lyon, France
marouane.fazouane@ensta.org, daniel.le-metayer@inria.fr
[2] Ulm University, Ulm, Germany
{henning.kopp,rens.vanderheijden,frank.kargl}@uni-ulm.de

Abstract. Electric vehicles are an up-and-coming technology that provides significant environmental benefits. A major challenge of these vehicles is their somewhat limited range, requiring the deployment of many charging stations. To effectively deliver electricity to vehicles and guarantee payment, a protocol was developed as part of the ISO 15118 standardization effort. A privacy-preserving variant of this protocol, POPCORN, has been proposed in recent work, claiming to provide significant privacy for the user, while maintaining functionality. In this paper, we outline our approach for the verification of privacy properties of the protocol. We provide a formal model of the expected privacy properties in the applied Pi-Calculus and use ProVerif to check them. We identify weaknesses in the protocol and suggest improvements to address them.

Keywords: privacy, formal verification, electric vehicle charging.

1 Introduction

In the current practice of charging for electric vehicles, the user of the vehicle typically has a contract that allows her to charge the vehicle at any compatible charging station. The contract is comparable to a mobile phone contract, as it enables roaming by users to charging stations operated by other companies, such as an electricity provider. However, the current standards for the implementation of such contracts, which should guarantee energy delivery and payment, do not fully consider the issue of location privacy for users. For example, a charging station operator could identify users from specific energy providers, which usually operate regionally. On the other hand, the company with which the user has a contract can track her movements through the different charging stations or energy providers involved.

One of the major challenges to design a protocol for electric vehicle charging is that the privacy of the user is difficult to express. One of the sources of this complexity is the potential overlap between the responsibilities of different participants of the protocol. Broadly speaking, two different approaches

F. Piessens et al. (Eds.): ESSoS 2015, LNCS 8978, pp. 17–33, 2015.

have been followed to express privacy requirements: the quantitative approach and the qualitative approach. The first category relies on the definition of appropriate "privacy metrics", such as k-anonymity, t-closeness [21], differential privacy [10,11] or entropy [25]. The second category consists in defining privacy properties in a suitable language endowed with a formal, mathematical semantics [32], and to use verification tools to reason about these properties. This approach is akin to the traditional analysis of security protocols [5]. In this paper, we take the second approach, and study the application of formal privacy analysis to protocols, using the POPCORN electric vehicle charging protocol [14] as a case study. This protocol is based on the current draft standard [15,16] and is thus much closer to practical application than earlier work on the privacy analysis of electric vehicle charging protocols [20]. Considering that privacy is a quite complex and multi-faceted notion, the first challenge in this context is the definition of appropriate privacy properties. In this paper, we define the privacy requirements of the protocol as a collection of six privacy properties and propose a definition of these properties in the applied Pi-Calculus [32]. Then we proceed with the analysis of the protocol and suggest several modifications to obtain an enhanced protocol meeting the required properties.

The contribution of the work presented in this paper is threefold:

– On the technical side: we provide the first realistic privacy compliant electric vehicle charging protocol improving the current draft of the ISO standard.
– On the formal side: we define privacy properties suitable for electric vehicle charging protocols including a new form of unlinkability (of the users and their uses).
– On the methodological side: we show how formal verification techniques can be applied in a "real-world" setting.

Beyond this specific protocol, this work also paves the way for a more general "privacy by re-design approach".

The remainder of the paper is structured as follows. Section 2 introduces POPCORN and discusses the relevant parts of the protocol and their relation to each other. Section 3 provides the formalization of POPCORN and discusses the relevant privacy properties. The results of our analysis are presented in Section 4, followed by suggestions of enhancements in Section 5. Related work is discussed in Section 6 and concluding remarks in Section 7.

2   Electric Vehicle Charging

In this paper we focus on a charging protocol for electric vehicles, which allows charging of a vehicle at a charging station in exchange for financial compensation. In practice, electricity is provided on a contract basis; therefore, financial compensation is implemented through contracts. Because energy providers often operate on a regional basis, roaming services are offered through a mobility operator. Thus, the following parties are involved in such a protocol:

Electric Vehicle (EV): This is the electric vehicle initiating the charging protocol.

Energy Provider (EP): The energy provider is the party providing the energy for recharging. This entity must receive compensation for its energy.

Charging Station (CS): The charging station is the device connecting the vehicle to the power grid to charge. In some scenarios, this may be operated by a charging station operator (CSO), although in practice this is usually the EP, who has complete control over the CS.

Mobility Operator (MO): The MO provides roaming contracts, so that the user can charge his vehicle with any EP covered by the contract. The MO has roaming agreements with one or more EPs and takes care of correct payments to these EPs as they are used by the users with whom the MO has contracts. In practice, some EPs also offer roaming contracts; thus, in some transactions, the EP and the MO may be the same entity.

This set of parties is also defined by the industry standard for electric vehicle charging, ISO 15118 [15,16]. The protocol for charging defined by this standard is designed in such a way that no party can cheat. However it does not provide any protection against user tracking. For example, the CS can recognize and distinguish EVs based on their identifiers, which are used to authenticate the vehicle and guarantee payments. Similarly, the MO could collect a list of visited CSs to track the trajectory of an EV. These location privacy issues were not sufficiently addressed in the standard. Several of these issues were identified by Höfer et al. [14], who subsequently proposed a more privacy-preserving protocol called POPCORN.

In POPCORN, several technical and organizational measures were added to improve the privacy of the charging protocol. The following parties were added:

Dispute Resolver (DR): The dispute resolver is a trusted third party, which comes into play only if one of the parties tries to cheat (for example, by not paying, or by claiming missing payment for transactions that were already paid). The DR can then resolve the identity of the cheating party and reconstruct the necessary transactions.

Payment Handler (PH): The payment handler is a trusted third party, similar to a bank, which handles the payment process between the MO and the EP. This party should ensure that the MO learns nothing about the EP. On the other hand, the EP should also not learn from which MO he is paid. The MO knows the identities of users, but not their locations. In contrast, the EP knows locations, but is not able to link them to individuals. The PH plays the role of an intermediary between those two partial datasets.

Although POPCORN does introduce additional privacy, as argued by Höfer et al., no formal privacy property was defined. The objective of this paper is precisely to address this issue and to challenge these privacy properties. Let us note also that the verification of integrity (non cheating) properties is not a goal of our analysis.

We now provide an outline of the different phases of the POPCORN protocol. A full specification can be found in [14].

Phase 0: Mobility Contract Establishment. First of all, the EV user signs a contract with some MO. The EV obtains anonymous contract credentials from a global certificate authority. This could be done, e.g., with Idemix [8], which makes it possible to hide the contract credentials from the global certificate authority to ensure privacy.

Furthermore group signature credentials are installed in the EV, where the group manager is the DR. These credentials allow the EV to sign messages. Any actor can then verify that the signatures belong to a member of the group but only the DR can reveal which EV exactly provided the signature.

Phase 1: Contract Authentication. When an EV is plugged into a CS it establishes a link over TLS. It then proceeds to prove that its contract and its anonymous credentials have not expired. The EV does not disclose any other contract information.

Phase 2: Charging Loop with Meter Receipts. The CS delivers energy to the EV and every time a fixed quantity has been transferred, the CS sends a meter reading. The EV then has to sign this meter reading with his group signature, thus committing to the reading. The CS checks the signature and, if it is correct, the charging loop continues until the EV is fully recharged or fails to produce a correct signature.

At the end of the charging loop, the CS sends to the EV a partial service Detail Record. This partial SDR contains the amount of electricity the EV has received, the payable amount and the recipient of the payment, i.e., the EP. Furthermore it contains an identifier, called the transaction id, which identifies the charging session represented by a specific SDR. The information about the EP in the SDR is encrypted with the public key of the PH.

Finally, the CS anonymously sends the group-signed commitments and the partial SDR to the EP.

Phase 3 and 4: SDR Delivery and Payment. After charging, the EV appends his probabilistically encrypted contract ID to the SDR, signs it, and forwards it to the MO. The MO can now extract the contract ID and thus the user but does not learn anything about the CS and the EP. He then sends the SDR together with the encrypted EP and the transaction number to the PH[1]. The payment Handler can then recover the EP and perform the payment. The mobility Operator obtains a receipt from the PH to confirm the payment.

Phase 5: Dispute Resolution. This phase is optional: it takes place only if the payment of the EP does not arrive withing the payment period. In this case, the EP forwards his partial SDR to the dispute Resolver who can then uncover the identity of the vehicle, since he is the group manager and the SDR was signed with the private key of the EV. He then informs the MO of the missing payment and requests the payment receipt. If the mobility Operator cannot provide the receipt, he has to settle the missing payment.

---

[1] If the EV tries to cheat here, the dispute resolution phase will allow the parties to determine the identity of the EV.

3   Formalization

In this section, we present a general framework for the formal description of cryptographic primitives and communication protocols. We then introduce privacy properties suitable for electric vehicle charging and show how to express them in this framework.

In this paper, we use the Dolev-Yao model, which is one of the standard models for analysing cryptographic protocols. The Dolev-Yao assumptions are the following: an open and unlimited network in which every agent can intercept, send and resend any message to any other agent. As a result, active adversaries can know every message exchanged in the network, and synthesize, modify, reroute, duplicate or erase messages or even impersonate any other agent. They can also use an algebraic system to derive new knowledge from the knowledge they already have. A minimal algebraic system makes it possible to create tuples and to apply projections to extract information from them. As shown in the next subsection, more powerful systems typically include cryptographic primitives and equational theories to reason about them. Even though it involves strong adversaries, the Dolev-Yao model requires additional assumptions on their computational capabilities:

- They cannot guess random numbers in sufficiently large state spaces.
- They cannot know more than what the algebraic system can prove. Typically, adversaries cannot extract the key or the plaintext from a ciphertext unless they already know the key (or the algebraic system includes rules to derive them).

Symbolic manipulations can be used to capture the properties of cryptographic primitives. Most model checkers and theorem provers used in this context abstract away from cryptography. However, results that are proven true by these tools are not necessarily sound with respect to a computational model. The computational soundness of the cryptographic primitives is needed to establish the soundness of the results. In essence, model checkers and theorem provers assume that the cryptographic primitives are secure; they only focus on proving that the *protocol* is secure, not the primitives on which it relies.

3.1   Modeling Cryptographic Primitives

The abstraction of cryptographic primitives is performed by using equational theories, which can be introduced by means of proof systems, as shown in the following. The syntax of terms is defined as follows:

$$M ::= id \mid f(M, ..., M) \mid M\{id/id\} \mid M\sigma$$

where id represents variable names and $f(M, ..., M)$ represents function application. Renaming is written $M\{id/id\}$, i.e., $f(x,y)\{z/y\}$ is equivalent to $f(x,z)$. Substitutions are written $M\sigma$ and can be used to substitute a whole term to a name: $f(x,y)\{g(t,x)/y\}$ is equivalent to $f(x, g(t,x))$.

We introduce two types of sequences: E is a sequence of equations over typed terms and Σ a sequence of constants and function signatures. We have the following system, which is a more natural reformulation of the equational theories discussed in [32]. The system can be easily extended with specific typing or well-formedness rules.

REFLEXIVITY

$$\overline{E,\Sigma \vdash M=M}$$

AXIOM

$$\frac{(M=N)\in E}{E,\Sigma \vdash M=N}$$

SYMMETRY

$$\frac{E,\Sigma \vdash M=N}{E,\Sigma \vdash N=M}$$

TRANSITIVITY

$$\frac{E,\Sigma \vdash M=N,\ E,\Sigma \vdash N=L}{E,\Sigma \vdash M=L}$$

APPLICATION

$$\frac{E,\Sigma \vdash M_i=N_i,\ E,\Sigma \vdash M_i:T_i,\ i=1..k,\ E,\Sigma \vdash f:T_1\times...\times T_k \to T}{E,\Sigma \vdash f(M_1,...,M_k)=f(N_1,...,N_k)}$$

SUBSTITUTION

$$\frac{E \vdash M=N}{E \vdash M\sigma=N\sigma}$$

RENAMING

$$\frac{E \vdash M=N}{E \vdash M\{m/n\}=N\{m/n\}}$$

If the theory E,Σ is decidable, we can define the congruence $=_{E,\Sigma}$ as $M =_{E,\Sigma} N \iff E,\Sigma \vdash M = N$. We can alternatively introduce this relation by means of reduction systems and we will require the system to be convergent, i.e., confluent and terminating. We will then have $M =_{E,\Sigma} N \iff \exists L, M \to^*_{E,\Sigma} L \wedge N \to^*_{E,\Sigma} L$.

Example. One way to model asymmetric encryption and digital signatures, uses the following sequences:

$\Sigma = [aenc : Pkey \times T \to T,\ adec : Skey \times T \to T,\ Pk : Skey \to Pkey,$
$\quad Sign : SKey \times T \to Signature,\ CheckSign : Pkey \times Signature \to Bool,$
$\quad RecoverData : Signature \to T,\ true : Bool]$

$E = [adec(x, aenc(Pk(x), y)) \to y,\ CheckSign(Pk(x), Sign(x, y)) \to true,$
$\quad RecoverData(Sign(x, y)) \to y]$

It is easy to prove that this system is convergent. For a more elaborate example, one may want to use automated tools like theorem provers together with a dedicated language to use equational theories or symbolic representation of cryptographic primitives.

3.2 Modeling Protocol Interactions and Concurrency

Process algebra provide convenient ways to formally describe high-level process interaction, communication and synchronization. They are also supported by useful tools for the formal analysis of processes. The applied Pi-Calculus is a language of this family and an extension of the well known Pi-Calculus and is the formal language that we use to model POPCORN. It has been introduced in [32]. An Applied Pi-Calculus process is defined by the following syntax:

$$P ::= 0 \mid phase\ n; P \mid \nu id : T;\ P \mid let\ id = M\ in\ P$$
$$\mid if\ M =_{E,\Sigma} M\ then\ P\ else\ P \mid in(id, id : T);\ P \mid out(id, M);\ P$$
$$\mid P|P \mid !P$$
$$M ::= id \mid f(M, \ldots, M) \mid Const \mid \ldots$$
$$Const ::= true \mid false \mid \ldots$$

Informally, its semantics is defined as follows:

Null Process: 0 is the null process. It does not reduce to any other process. This is often omitted after an $in()$ or an $out()$, i.e., write $out(x, y)$ instead of $out(x, y); 0$.

Phase: $phase\ n; P$ is a synchonization point, i.e., all instructions from the previous phase are discarded before starting phase n and it behaves as P. By default, processes run in *phase* 0. In particular, this will be useful to model offline attacks.

Restriction: $\nu x : T;\ P$ creates a new typed name x, adds it to the scope of P and then behaves as P.

Let Definition: $let\ x = M\ in\ P$ binds the value of the term x to the term M; then it behaves as P.

IF-THEN-ELSE Statement: $if\ M_1 =_{E,\Sigma} M_2\ then\ P_1\ else\ P_2$ reduces to P_1 if $M_1 =_{E,\Sigma} M_2$ is provably true in the equation theory E, Σ. It reduces to P_2 otherwise (not only when $M_1 =_{E,\Sigma} M_2$ is false, but also when it is not provable).

Input: $in(id_1, id_2 : T);\ P$ waits on channel id_1 for a term of type T to be output, reads its value into the term id_2 and then behaves like P.

Output: $out(id, M);\ P$ waits for another process to listen to the channel id (i.e. a process of the form $in(id, id_2);\ P_2$ that runs in parallel). It can then output the term M in the channel id and behaves like P.

Parallel Composition: $P_1|P_2$ basically means that both P_1 and P_2 run in parallel.

Replication: $!P$ represents an unbounded number of parallel replications of the process P.

The Pi-Calculus and the Applied Pi-Calculus have formal semantics defined by labelled transition systems. The weak bisimilarity $\approx$ is proven to coincide with the observational semantics of the language [30,32]. This gives us a mechanical way to prove observational semantics. The latter basically means that two processes are equivalent if and only if no static context (restriction of names and parallel composition with an arbitrary process) can distinguish between the two processes. This gives us the ability to express some advanced privacy-related requirements. However, $\approx$ is undecidable and some advanced methods are needed to prove observational equivalence. For instance, ProVerif, the automated tool used in our analysis, applies specific heuristics to prove a stronger notion than observational equivalence which is decidable. For this reason, ProVerif fails to prove some valid equivalences.

3.3   Privacy-Related Properties

We shall present in this subsection the main privacy properties we are interested in. We shall then show how to express and verify each of them in the Applied Pi-Calculus. The first five properties have been studied in different papers. As far as we know, this is not the case for the last property. The properties are the following:

Weak Secrecy/Confidentiality: Active adversaries cannot deduce the complete secret from their interactions with the communicating parties. Or, equivalently, adversaries cannot output the secret on a public channel [32]. This property is interesting in the context of POPCORN to express the fact that the identity of the vehicles should remain secret as well as the identities of the mobility Operator (MO) and the energy Provider (EP) communicating with a specific vehicle.

Strong Secrecy: This is a stronger notion where adversaries cannot even distinguish if the secret changes [32]. It provides stronger guarantees on the confidentiality of the secret, excluding any partial knowledge of the secret. In contrast, *Weak Secrecy* is not breached as long as the adversary cannot deduce the complete secret.

Anonymity: A particular user may use the service without others noticing him. This means, informally, that he can use the service without disclosing personally identifiable information during this process [7]. This property is useful in a scenario where the identity of the user (or vehicle) is known to the adversary, which can be, for example, a Charging Station (CS) knowing the identity of the vehicle from a previous interaction through a non-privacy preserving protocol. In this particular scenario, the adversary should not know that a specific vehicle has used the service.

Resistance to Offline-Guessing Attacks: Adversaries cannot distinguish between correct and incorrect guesses [6].

Strong Unlinkability: A user may make multiple uses of a service without others being able to establish a link between them [7]. This property guarantees that users cannot be traced by active adversaries like CSs and EPs.

Unlinkability of uses and users: A user may use the service without others being able to link him to a particular use. This property is not to be confused with *Strong Unlinkability*. It guarantees that an active adversary, like an MO who already knows sensitive data, should not be able to link the known vehicle to a complete bill which contains charging sessions' details and metadata.

In the case of POPCORN we are interested in the first four properties for the identities of the electric Vehicle (EV), its MO and the CS/EP it may communicate with. *Strong Unlinkability* is only studied with respect to the identity of the user (or EV). The last property will be studied for a particular user and a particular usage.

The above properties can be expressed as follows in the formal model:

Weak Secrecy/Confidentiality: This property can be expressed as a reachability property [32].

Strong Secrecy: The unobservability of secret changes can be captured using observational equivalence [32].

$$P\{M_1/secret_1, \ldots \ M_n/secret_n\} \approx P\{M_1'/secret_1, \ldots \ M_n'/secret_n\}$$

The above formula states that it is possible to replace secrets by different values in the protocol without active adversaries being able to distinguish these situations, which is exactly *Strong Secrecy*.

Anonymity: The fact that a particular user can remain unnoticed can be expressed as: [7]

$$!(\nu id; \ P)| \ P\{M/id\} \approx !(\nu id; \ P)$$

Resistance to Offline-Guessing Attacks: We can encode guesses by outputting the correct or a dummy value on a public channel. The property is true if and only of the following holds [6]:

$$P| \ (phase \ 1; \ out(publicChannel, \ secret)) \approx$$
$$P| \ (phase \ 1; \ \nu dummy. \ out(publicChannel, \ dummy))$$

Strong Unlinkability: This property amounts to checking whether the protocol is equivalent to a version of itself in which the number of sessions is limited to one per user [7].

$$!(\nu id; \ !P) \approx !(\nu id; \ P)$$

Unlinkability of Uses and Users: This property expresses the fact that is a transaction cannot be linked to a given user. Typically, in the case of POPCORN, we have two bills per charging session: one for the MO and one for the EP. If an adversary MO knows the EV and both bills, verifying the property is equivalent to answering the following question: can this adversary link a bill he knows with the bill containing the detailed charging information of the vehicle? The property is expressed as follows:

$$P_1 \approx P_2 \text{ where}$$

- $P_1 = C[phase \ 1; \ (out(publicChannel, \ trid\_CS)| \ out(publicChannel, \ trid\_EV))]$
- $P_2 = C[phase \ 1; \ (out(publicChannel, \ dummy_1)| \ out(publicChannel, \ dummy_2))]$
- $(trid\_EV, trid\_CS)$ denotes a charging session where $trid\_EV$ is the transaction id on the user side and $trid\_CS$ is the transaction id on the charging station side.
- $dummy_1$ and $dummy_2$ are two valid but unlinked charging session identifications.
- $C[]$ an arbitrary context such that $C[0]$ represents the studied protocol.

As an illustration, a typical adversary would have the following template, where $link()$ encodes strategies that can link both transactions (in ProVerif syntax):

```
let Adversary(...)=
  phase 1;
  in(publicCh,trid1:TransactID);
  in(publicCh,trid2:TransactID);
  if link(trid1,trid2,extra_infos)=true then
```

```
(* successfully  linked  the  usage  to  the  user*)
new message: MSG;
(* 'c' is  public  and  not  used  elsewhere*)
out(c,message)
```

4   Verification of Privacy Policies

The first step of our methodology consists in translating the informal description
of the protocol (which may be unclear or incomplete) to simple diagrams including a complete description of each step. This representation is then translated
into the Applied Pi-Calculus. The next step is the definition of the privacy properties following the approach described in the previous section. These properties
are then submitted to ProVerif for verification. To this aim, we have implemented
a convergent equational theory that captures all cryptographic primitives required by POPCORN. The properties that cannot be verified by ProVerif can
either be shown to be incorrect or proven by hand. The failure to prove a correct
property can be due either to the limitation of the tool described in Section 3.2
or because of inappropriate design choices for the model of the protocol.

4.1   Minor Problems and Adjustments

It is possible to exploit the signature of the meter readings to generate adversaries
that can invalidate *Strong Unlinkability*. Indeed, a malicious CS that generates
twice the same meter reading in two different sessions obtains the same signed
value if and only if the EV is the same in both session (and thus iff *Strong
Unlinkability* is not satisfied). We can easily confirm this claim by submitting
the following equivalence to ProVerif:

```
free gmsk: gmskey [private].  (* master key *)
equivalence
(  (* Multiple  sessions  *)
  !( new id: ID;
    !(let gsk=GKeygen(gmsk,id) in
      in(publicChannel,m: bitstring);
      out(publicChannel,(GPk(gmsk),Sign(gsk, m))) ) )
)
(  (* Single  session  *)
  !( new id: ID;
    (let gsk=GKeygen(gmsk,id) in
      in(publicChannel,m: bitstring);
      out(publicChannel,(GPk(gmsk),Sign(gsk, m))) ) )
)
```

Automated payment gives rise to another weakness. In fact, according to the
ISO/IEC 15118 specification, the EP's identification number is included in the

bill that the EV sends to the MO. In contrast, in POPCORN, an encrypted identification of the energy provider is sent to the payment Handler (PH). However,
the POPCORN description was imprecise about how this actually works. If a
standard asymmetric encryption is used, the problem again is that even though
an adversary cannot find the secret, he can still detect its changes, which means
that *Strong Secrecy* is not satisfied. We can confirm this claim using the following
ProVerif program, which attempts to verify observational equivalence:

```
free sk: skey [private].
free id: bitstring [private]. (* the secret *)
noninterf id. (* strong secrecy *)
process
   out(publicChannel ,(Pk(sk),aenc(Pk(sk),id)))
```

Possible fixes: The above problems can be fixed as follows. First, a session
number can be added to the data to be signed: $Sign(gsk, (session\_id, m))$.
Signatures are now cryptographically linked to a specific session and cannot be
used in two different sessions by a malicious CS/EP. For the encrypted identity
of the EP an option is to use randomized encryption:

$$\nu r : nonce; \ aenc(Pk(sk\_PH), (r, idEP))$$

4.2   Results Using ProVerif

We consider now that the changes presented in the previous subsection have
been added to POPCORN. The analysis of the protocol using ProVerif returns
the following results:

	true	cannot be proven
Weak Secrecy (*EV,MO*)	✓	
Strong Secrecy (*EV,MO*)	✓	
Resistance to Offline-Guessing Attacks (*EV,MO*)	✓	
Anonymity (*EV*)		✓
Strong Unlinkability (*EV*)		✓
Weak Secrecy (*EP*)		✓

4.3   Unlinkability of Uses and Users

In this subsection we show that the remaining property, *Unlinkability of uses and
users*, does not hold. To do so, we will exploit two different aspects of POPCORN
and prove that minor changes can lead either to a broken protocol or to a variant
of POPCORN that does not verify this property. Therefore, more substantial
changes are needed, which are discussed in the next section.

Exploiting Automated Payment. Since transaction numbers are contained in the bills obtained by EPs and MOs, an adversary can simply compare the two values to link them. The linking function of the typical adversary presented in Section 3.3 would be:

```
letfun link(trid1 : TransactID , trid2 : TransactID)=
   trid1=trid2 .
```

It is easy to verify that $P_1 \not\approx P_2$ in this case, because a message is output in channel c in the first process and not in the second one. Thus, *Unlinkability of uses and users* is not satisfied. We must find a way to generate two transaction identifiers that can only be linked by the actor generating them or by the PH. The PH should also have the ability to derive one transaction number from the other one.

Exploiting Dispute Resolution. We consider at this point that all minor modifications suggested above have been added to POPCORN. During dispute resolution, as explained in Section 2, the EP contacts the dispute Resolver (DR) with the unpaid bill. Then the latter unveils the identity of the vehicle and contacts its MO with the transaction number of the unpaid charging session. Since dispute resolution must be functional, the MO can verify that the EP-side transaction number is linked to one of the paid or unpaid sessions. The MO has the ability to link two transaction numbers. The linking function in that case would be a function corresponding to the procedure used by the MO. Thus, even with these minor modifications, *Unlinkability of uses and users* is still not satisfied. A remaining option consists in modifying dispute resolution, which is discussed in the next section.

5   Remedy

The idea behind the suggested remedy is to involve the PH in *Dispute Resolution*. To implement this solution, we need a transaction id scheme that can ensure unlinkability and unforgeability.

create(pk_PH, rand, token): this randomized constructor returns *(transactID_user, transactID_server, pi)* such that *transactID_user* and *transactID_server* are two transaction numbers that can only be linked by the PH and *pi* is a Zero Knowledge Proof such that
$VerifyProof(pk\_PH', token', transactID\_server, pi) = true$ iff
$(token, pk\_PH) = (token', pk\_PH')$.
check(pk_PH,token, transactID_server, pi) returns *true* iff
$\exists rand.\exists transactID\_user.$
$(transactID\_user, transactID\_server, pi) = create(pk\_PH, rand, token).$
getUserSideTransactionID(sk_PH,transactID_server) returns
transactID_user such that $\exists rand.\exists token.\exists pi.$
$(transactID\_user, transactID\_server, pi)$
$= create(Pk(sk\_PH), rand, token).$

getServerSideTransactionID(sk_PH,transactID_user) returns
  transactID_server such that $\exists rand.\exists token.\exists pi.$
  (transactID_user, transactID_server, pi)
  $= create(Pk(sk\_PH),\ rand,\ token).$

Some changes have to be made to the protocol to use the above cryptographic primitives; the corresponding diagrams can be found in Appendix A.

Transaction Numbers Establishment: the CS chooses the *token* and sends it to the EV through a secure channel. The latter chooses a random nonce r and uses it to generate the transaction ids: *(trid_EV, trid_CS, pi)* $=$ *create(pk_PH, r, token)*. The electric Vehicle then sends *trid_CS* and *pi* to the charging Station. The charging Station checks the validity of the transaction number before using it: *check(pk_PH, token, trid_CS, pi)*.

Automated Payment: upon receiving a payment order from an MO, the PH computes the id of the EP that should be contacted but also the correct transaction number that should be paid for:
trid_CS $=$ getServerSideTransactionID(sk_PH, trid_EV). *trid_EV* being the transaction id on the mobility Operator side.

Dispute Resolution: during dispute resolution, the DR should contact the PH with an unpaid *trid_CS* and the identity of the MO that should be contacted. The PH will compute the correct transaction id for which the MO should pay: *trid_EV $=$ getUserSideTransactionID(sk_PH, trid_CS)*.

6   Related Work

The definition of appropriate frameworks to express and reason about privacy properties has generated a significant interest over the last decade. Indeed, privacy is a complex and subtle notion, and the first challenge in this area is defining formal properties that reflect the actual privacy requirements. A variety of languages and logics have been proposed to express privacy policies [2,3,4,18,17,23,27,35]. These languages may target citizens, businesses or organizations. They can be used to express individual privacy policies, corporate rules or legal rules. Some of them make it possible to verify consistency properties or system compliance. For example, one may check that an individual privacy policy fits with the policy of a website, or that the website policy complies with the corporate policy. These verifications can be performed either *a priori*, on the fly, or *a posteriori*, using techniques like static verification, monitoring and audits. Similarly, process calculi like the applied Pi-Calculus have already been applied to define and verify privacy protocols [9]. Process calculi are general frameworks to model concurrent systems. They are more powerful and versatile than dedicated frameworks, which is illustrated in this work. The downside is that specifying a protocol and its expected properties is more complex. To address this issue, some authors propose to specify privacy properties at the level of architectures [1,19]. For example, the framework introduced in [19] includes an inference system to reason about potential conflicts between confidentiality and accountability requirements. Other approaches are based on deontic logics, e.g. [12], which focuses on expressing policies and their relation to database security or distributed systems. A difficulty with epistemic logics in this context is the problem known as "logical omniscience". Several ways to solve this difficulty have been proposed [13,31].

Privacy metrics such as k-anonymity [22,33], l-diversity [26] or ϵ-differential privacy [10,11] have also been proposed as ways to measure the level of privacy provided by an algorithm. Differential privacy provides strong privacy guarantees independently of the background knowledge of the adversary. The main idea behind ϵ-differential privacy is that the presence or absence of an item in a database should not change in a significant way the probability of obtaining a certain answer for a given query. Methods [11,28,29] have been proposed to design algorithms meeting these privacy metrics or to verify that a system achieves a given level of privacy [34]. These contributions on privacy metrics are complementary to our work, as we follow a logical approach here, proving that a given privacy property is met (or not) by a protocol.

Liu et. al. [20] define a formal model for an electric vehicle charing protocol, differing in several ways from POPCORN. First, they do not distinguish between the MO and EP stakeholders, and they do not have a dedicated PH. Therefore they have only three parties: the user, the supplier, which in POPCORN is called the EV, and the judging authority, which is comparable to DR in POPCORN. Their protocol [24] also supports additional functionalities such as traceability (if the car is stolen), which is not proven to be privacy preserving and discharging of the EV, i.e., the EV can choose to sell energy back into the grid.

7   Conclusions

This paper presents an application of a formal approach to define a real life protocol meeting privacy requirements. Our formal model has made it possible to identify weaknesses in the original POPCORN protocol [14] and to suggest improvements to address these issues. POPCORN preserves the confidentiality of its users (*Weak Secrecy*) but *Strong Secrecy* and *Strong Unlinkability* are not satisfied by the original version of the protocol. However, minor modifications of the protocol are sufficient to redress these weaknesses. We have also shown that POPCORN does not ensure a particular form of unlinkability: it does not prevent an attacker from linking a user to his uses of the system. We have also argued that more significant changes in the definition of POPCORN are necessary to address this issue. The mitigation proposed here does not affect the functionality of the protocol and can be shown to meet the expected unlinkability property.

The work described in this paper can be seen as a contribution to privacy re-engineering which is of prime importance to enhance legacy systems to deal with privacy requirements. The next step in this direction would be to go beyond this specific protocol and provide a framework for privacy re-engineering. We believe that the re-design approach presented in [14] in association with the formal approach described here pave the way for the definition of an iterative improvement methodology that could form the core for such a framework. We would also like to stress that this approach should not be opposed to the "privacy by design" philosophy. Indeed, privacy requirements very often are (or seem to be) in conflict with other (functional or non functional) requirements. The iterative methodology suggested here could be applied at the level of specifications and seen as a strategy to address the needs on privacy by design in some situations.

References

1. Antignac, T., Le Métayer, D.: Privacy by Design: From Technologies to Architectures. In: Preneel, B., Ikonomou, D. (eds.) APF 2014. LNCS, vol. 8450, pp. 1–17. Springer, Heidelberg (2014)
2. Backes, M., Dürmuth, M., Karjoth, G.: Unification in Privacy Policy Evaluation - Translating EPAL into Prolog. In: POLICY, pp. 185–188 (2004)
3. Barth, A., Mitchell, J.C., Datta, A., Sundaram, S.: Privacy and Utility in Business Processes. In: CSF, pp. 279–294 (2007)
4. Becker, M.Y., Malkis, A., Bussard, L.: A Practical Generic Privacy Language. In: Jha, S., Mathuria, A. (eds.) ICISS 2010. LNCS, vol. 6503, pp. 125–139. Springer, Heidelberg (2010)
5. Blanchet, B., Abadi, M., Fournet, C.: Automated verification of selected equivalences for security protocols. In: Proceedings of the 20th Annual IEEE Symposium on Logic in Computer Science, LICS 2005, pp. 331–340. IEEE (2005)
6. Blanchet, B., Smyth, B.: Proverif 1.85: Automatic cryptographic protocol verifier, user manual and tutorial (2011)
7. Brusó, M., Chatzikokolakis, K., Etalle, S., den Hartog, J.: Linking Unlinkability. In: Palamidessi, C., Ryan, M.D. (eds.) TGC 2012. LNCS, vol. 8191, pp. 129–144. Springer, Heidelberg (2013)
8. Camenisch, J., Van Herreweghen, E.: Design and implementation of the idemix anonymous credential system. In: Proceedings of the 9th ACM Conference on Computer and Communications Security, pp. 21–30. ACM (2002)
9. Delaune, S., Kremer, S., Ryan, M.D.: Verifying Privacy-type Properties of Electronic Voting Protocols. Journal of Computer Security 17(4), 435–487 (2009), http://www.lsv.ens-cachan.fr/Publis/PAPERS/PDF/DKR-jcs08.pdf
10. Dwork, C.: Differential Privacy. In: Bugliesi, M., Preneel, B., Sassone, V., Wegener, I. (eds.) ICALP 2006. LNCS, vol. 4052, pp. 1–12. Springer, Heidelberg (2006)
11. Dwork, C.: A firm foundation for private data analysis. Commun. ACM 54(1), 86–95 (2011)
12. Glasgow, J., MacEwen, G., Panangaden, P.: A logic for reasoning about security. In: Proc. of the 3rd Computer Security Foundations Workshop, pp. 2–13 (1990)
13. Halpern, J.Y., Pucella, R.: Dealing with Logical Omniscience. In: Proc. of the 11th Conf. on Th. Aspects of Rationality and Knowl., pp. 169–176. ACM, USA (2007), http://doi.acm.org/10.1145/1324249.1324273
14. Höfer, C., Petit, J., Schmidt, R., Kargl, F.: POPCORN: privacy-preserving charging for eMobility. In: Proceedings of the 2013 ACM Workshop on Security, Privacy & Dependability for Cyber Vehicles, pp. 37–48. ACM (2013)
15. ISO: Road vehicles - Vehicle-to-Grid Communication Interface - Part 1: General information and use-case definition. ISO 15118, International Organization for Standardization, Geneva, Switzerland (2012)

16. ISO: Road vehicles - Vehicle-to-Grid Communication Interface - Part 2: Technical protocol description and Open Systems Interconnections (OSI) layer requirements. ISO 15118, International Organization for Standardization, Geneva, Switzerland (2012)
17. Jafari, M., Fong, P.W.L., Safavi-Naini, R., Barker, K., Sheppard, N.P.: Towards defining semantic foundations for purpose-based privacy policies. In: CODASPY, pp. 213–224 (2011)
18. Le Métayer, D.: A Formal Privacy Management Framework. In: Degano, P., Guttman, J., Martinelli, F. (eds.) FAST 2008. LNCS, vol. 5491, pp. 162–176. Springer, Heidelberg (2009)
19. Le Métayer, D.: Privacy by Design: A Formal Framework for the Analysis of Architectural Choices. In: Proc. of the 3rd ACM Conference on Data and Application Security and Privacy, pp. 95–104. ACM, USA (2013), http://doi.acm.org/10.1145/2435349.2435361
20. Li, L., Pang, J., Liu, Y., Sun, J., Dong, J.S.: Symbolic analysis of an electric vehicle charging protocol. In: Proc. 19th IEEE Conference on Engineering of Complex Computer Systems (ICECCS 2014). IEEE Computer Society (2014)
21. Li, N., Li, T., Venkatasubramanian, S.: t-Closeness: Privacy Beyond k-Anonymity and l-Diversity. In: IEEE 23rd International Conference on Data Engineering, pp. 106–115 (April 2007)
22. Li, N., Qardaji, W.H., Su, D.: Provably Private Data Anonymization: Or, k-Anonymity Meets Differential Privacy. CoRR abs/1101.2604 (2011)
23. Li, N., Yu, T., Antón, A.I.: A semantics based approach to privacy languages. Comput. Syst. Sci. Eng. 21(5) (2006)
24. Liu, J.K., Au, M.H., Susilo, W., Zhou, J.: Enhancing location privacy for electric vehicles (at the *right* time). In: Foresti, S., Yung, M., Martinelli, F. (eds.) ESORICS 2012. LNCS, vol. 7459, pp. 397–414. Springer, Heidelberg (2012)
25. Ma, Z., Kargl, F., Weber, M.: A location privacy metric for V2X communication systems. In: IEEE Sarnoff Symposium, pp. 1–6 (March 2009)
26. Machanavajjhala, A., Gehrke, J., Kifer, D., Venkitasubramaniam, M.: l-Diversity: Privacy Beyond k-Anonymity. In: ICDE, p. 24 (2006)
27. May, M.J., Gunter, C.A., Lee, I.: Privacy APIs: Access Control Techniques to Analyze and Verify Legal Privacy Policies. In: CSFW, pp. 85–97 (2006)
28. McSherry, F.: Privacy integrated queries: an extensible platform for privacy-preserving data analysis. Commun. ACM 53(9), 89–97 (2010)
29. McSherry, F., Talwar, K.: Mechanism Design via Differential Privacy. In: FOCS, pp. 94–103 (2007)
30. Milner, R.: Communicating and Mobile Systems: The Pi-calculus. Cambridge University Press, New York (1999)
31. Pucella, R.: Deductive Algorithmic Knowledge. CoRR cs.AI/0405038 (2004)
32. Ryan, M.D., Smyth, B.: Applied pi calculus. In: Cortier, V., Kremer, S. (eds.) Formal Models and Techniques for Analyzing Security Protocols, ch. 6. IOS Press (2011), http://www.bensmyth.com/files/Smyth10-applied-pi-calculus.pdf
33. Sweeney, L.: k-Anonymity: A Model for Protecting Privacy. International Journal of Uncertainty, Fuzziness and Knowledge-Based Systems 10(5), 557–570 (2002)
34. Tschantz, M.C., Kaynar, D.K., Datta, A.: Formal Verification of Differential Privacy for Interactive Systems. CoRR abs/1101.2819 (2011)
35. Yu, T., Li, N., Antón, A.I.: A formal semantics for P3P. In: SWS, pp. 1–8 (2004)

A   POPCORN v2

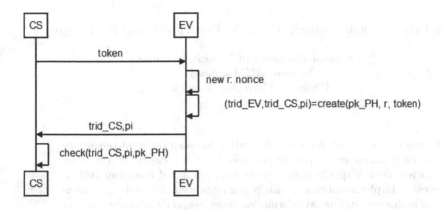

Fig. 1. Transaction number establishment

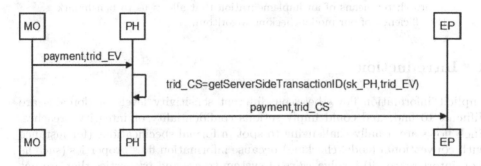

Fig. 2. Changes in automated payment

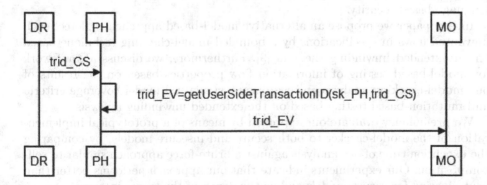

Fig. 3. Changes in dispute resolution

Idea: Unwinding Based Model-Checking and Testing for Non-Interference on EFSMs

Martín Ochoa[1], Jorge Cuéllar[2], Alexander Pretschner[1], and Per Hallgren[3]

[1] Technical University of Munich, Germany
[2] Siemens AG, Germany
[3] Chalmers University, Sweden

Abstract. Undesired flows of information between different sensitivity levels or domains can seriously compromise the security of a system. Moreover, even if specifications are secure, unwanted flows can still be present in implementations. In this paper we present a model-based technique to discover unwanted information flows in specifications and to test systems for unwanted flows. We base our approach on an unwinding relation for Extended Finite State Machines. We preliminary validate our approach by means of an implementation that allows us to benchmark the efficiency of our model-checking algorithm.

1 Introduction

Implicit information flows between different sensitivity levels or domains are difficult to find, and could imply serious confidentiality or integrity breaches. Such flows are usually challenging to spot in formal specifications (for instance with conventional model-checking) because information flow properties (such as Non-interference [6]) involve *pairs* of system traces and properties that are not expressible in conventional temporal logics. A way to cope with this problem is *self-composition* originally proposed by Barthe et. al. [1], in the context of language-based security.

In this paper we propose an alternative model-based approach to detect such unwanted flows in specifications by a bounded model-checking technique based on an extended unwinding theorem [9]. Furthermore, we discuss a framework for model-based testing of information flow properties based on the results of our model-checking analysis and we discuss structural model coverage criteria and mutation-based testing based on the extended unwinding analysis.

We preliminary evaluate our approach by means of a prototypical implementation of the model-checker to both secure and insecure models, by comparing the execution time of our analysis against a brute-force approach similar to self-composition. Our experiments indicate that our approach performs better than brute-forcing for secure models and as the depth of the traces increases.

The rest of the paper is organized as follows: Sect. 2 discusses some preliminaries about non-interference and sets the notation for the rest of the paper. Sect. 3 presents the approach for model-checking EFSMs based on an extended

F. Piessens et al. (Eds.): ESSoS 2015, LNCS 8978, pp. 34–42, 2015.

unwinding theorem. In Sect. 4 we show how to model-based test implementations for non-interference. In Sect. 5 we present the results of our evaluation experiments. Sect. 6 discusses related work and we conclude in Sect. 7.

2 Preliminaries

Non-interference and EFSMs The notion of information flow security is that an observer belonging to some security domain should not obtain information about (or be influenced by) actions happening in some other security domain. The most common instance is the two-element lattice $(\{L, H\}, \leq)$ where $\mathcal{I} = \mathcal{H}_I \uplus \mathcal{L}_I$ is partitioned into *low* and *high* inputs (and $\mathcal{O} = \mathcal{H}_O \uplus \mathcal{L}_O$ for the outputs) and information is allowed to flow from low (L) to high (H) but not vice versa.

Let $[\,] : \mathcal{I}^* \to \mathcal{O}^*$ be a function that takes sequences of input events in $\mathcal{I}$ into sequences of output events in $\mathcal{O}$. $\cdot|_L$ is a purging function that deletes elements in $\mathcal{H}_I$ from an input sequence (and elements in $\mathcal{H}_O$ from an output sequence respectively). Non-interference [6] is the property:

$$\forall \, i_1, i_2 \quad i_1|_L = i_2|_L \Rightarrow [i_1]|_L = [i_2]|_L \tag{1}$$

Definition 1. *An Extended Finite State Machine (EFSM) is a tuple*

$$M = (\mathbf{S}, S_0, \mathcal{V}, s_0, \mathbf{I} = \mathbf{H} \uplus \mathbf{L}, \mathbf{O}, \mathbf{T})$$

where $\mathbf{S}$ *is a finite set of states,* $S_0 \in S$ *is the* initial state, $\mathcal{V}$ *is a finite set of variables, which for simplicity we assume to be the same for all states* $S \in \mathbf{S}$, $s_0 : \mathcal{V} \to Val$ *is the* initial valuation *of the variables,* $\mathbf{I}$ *is the set of (parametrized) input symbols where* $\mathbf{H}$ *is the set of high inputs,* $\mathbf{L}$ *is the set of low inputs, and* $\mathbf{O}$ *is the set of (parametrized) output symbols, and* $\mathbf{T}$ *is a finite set of transitions* $\{T_1, T_2, T_3, \ldots, T_n\}$. *Transitions* $T_i \in \mathbf{T}$ *have the form:*

$$(S_1, S_2, I(\alpha), [C(\alpha, \overrightarrow{x})], \; \overrightarrow{x} := E(\alpha, \overrightarrow{x}), \; O(F(\alpha, \overrightarrow{x})))$$

where $\overrightarrow{x}$ *represents the variables of the state* S_1, *the symbol* α *represents the parameter of the input event* I, *the condition* C *is a decidable condition over* $\overrightarrow{x}$ *and* α, *the expression* E *is an expression to be assigned to the variable* x, *and the expression* F *is the parameter of the output event* O.

A natural semantics is associated with EFSMs, where input symbols trigger an immediate output and a transition in the machine, according to the value of the internal variables and the current state.

2.1 Unwinding-Based Verification

In [9] an unwinding theorem for EFSMs is presented that soundly approximates non-interference. Similarly to the original Unwinding Theorem [7], this extension constructs a relation between symbolic states and checks for consistency of the low outputs of related states (output consistency).

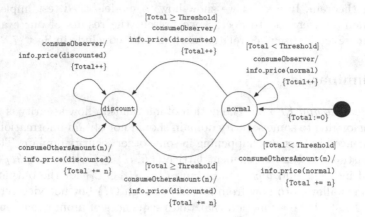

Fig. 1. Discounted consumption

The algorithm for approximating non-interference of [9] can be summarized as follows: **1.** First, compute a minimal unwinding relation of states : two states S_1 and S_2 are in relation if there is a transition triggered by a high input connecting them (local consistency) or if they are the goal states of two related states via the same low input (step consistency). **2.** Compute a set $T(S)$ of tainted variables for each state S where a variable can be tainted if its value is changed as a side effect of a high transition, or transitively in a transition guarded by a condition reading from a tainted variable. **3.** Check for output consistency of states in a relation by comparing the outputs of low inputs where the partition of the abstracted state via the guards is considered: if guards are tainted, all outputs must be equal regardless of the condition, otherwise syntactically identical guards are compared pair-wise.

Example. There is a number of consumers of some commodity. When the total consumption reaches a certain threshold, the price per unit is discounted. Consumers should not learn about the consumption of the others. This is modelled in Fig. 1, and formalized by establishing that `consumeObserver` is a low input, whereas `consumeOthersAmount` is a parametric high input. Note that the state `normal` is not output consistent, since for the low input `consumeObserver` guarded by the tainted variable `Total`, there are two different possible outputs (`info_price(normal)` and `info_price(discount)`). But can we find two concrete traces exhibiting a violation to the security property?

3   Unwinding Based Model Checking

Using the verification results of [9] it is possible to determine whether a specification is secure. Nevertheless, if a model is considered as flawed, it can be a false positive, given that the verification algorithm is sound but not complete. It is in general not trivial to find traces witnessing this violation. In this section we discuss an approach to generate such counterexamples by using the results of the unwinding analysis.

3.1 Trace Generation

To begin with, note that if it is not possible to construct a (minimal) unwinding relation, then there exists at least a pair of states (S, S') such that for some low input the specified outputs are inconsistent. We want to take advantage of this fact to produce two traces that are input equivalent, but such that they are not output equivalent for S and S'.

We say that a sequence of transitions $\mathcal{T} = \mathcal{T}_1 \mathcal{T}_2 \mathcal{T}_3 \ldots$ is an *abstract trace* for the symbolic Mealy machine, if $\mathcal{T}_0$ starts in S_0 and that $\mathcal{T}_i$ starts in the state in which $\mathcal{T}_{i-1}$ ends, for $i > 0$. Two abstract traces $\mathcal{T}_1 = \mathcal{T}_{1,1} \mathcal{T}_{1,2} \mathcal{T}_{1,3} \ldots$ and $\mathcal{T}_2 = \mathcal{T}_{2,1} \mathcal{T}_{2,2} \mathcal{T}_{2,3} \ldots$ are said to be low-input-equivalent modulo a set $\mathfrak{C}$ of constraints on the parameter symbols $\alpha_{i,j}$ of the transitions, written $\mathcal{T}_1 \overset{i}{\equiv}_L \mathcal{T}_2 \ (mod \ \mathfrak{C})$ if the constraint $\mathfrak{C}$ relates the parameters of low inputs in both abstract traces in a way that any concretization (variables with a concrete valuation) of the abstract traces is low-input-equivalent.

One can then consider all abstract traces conducting from the initial state to S and to S' respectively, up to a certain length k. From the cartesian product of those two sets, we compute all possible pairs of abstract traces such that $\mathcal{T}_1 \overset{i}{\equiv}_L \mathcal{T}_2$ and the associated set of constraints $\mathfrak{C}$ as defined by the guards and side effects of the model. At this point, we have candidates for traces that are abstractly low equivalent. We now automatically check for satisfiability which will provide witnesses for the values of the variables and input parameters at each state thus allowing to produce a fully instantiated pair of traces (if any).

Algorithm 1. Counterexample generation

```
 1: procedure COUNTEREXAMPLES(S, S')
 2:     C ← ∅
 3:     T ← FINDPATHS(S_0, S, k)
 4:     T' ← FINDPATHS(S_0, S', k)
 5:     for all (𝒯, 𝒯') ∈ T × T' do
 6:         if INPEQUIVALENT(𝒯, 𝒯') then
 7:             C ← C ∪ (𝒯, 𝒯')
 8:         end if
 9:     end for
10:     C' ← SAT(C)
11:     return C'
12: end procedure
```

Formally, we use the Algorithm 1, where FINDPATHS(S_0, S, k) returns all abstract traces between S_0 and S of length at most k. The function INPEQUIVALENT compares two abstract traces and determines whether they are low input equivalent. Finally, SAT(C) returns a pair of concrete traces for each satisfiable low equivalent abstract pair and its associated set of constraints $\mathfrak{C}$ as defined by the model, plus the condition that the last output must differ, which will imply a violation to non-interference.

Example. Consider the EFSM in Fig. 1. As discussed before, it is not output consistent for $S = S' = $ normal. We now apply Algorithm 1. It suffices to find all paths of size $k = 1$ to normal. We extend these traces with the non output-consistent transitions of S (S'), in this case the two transitions triggered by consumeObserver. We set Threshold=10 and run the constraint solver, obtaining the pair of traces: {(consumeOthersAmount(0) consumeObserver, consumeOthersAmount(11) consumeObserver)} which constitute a counterexample, since they produce two different observable outputs for low.

4   Model-Based Non-interference Testing

In this section we discuss model-based testing strategies for non-interference. As it is the case with testing in general, it is computationally infeasible to exhaustively test the SUT. It is hence necessary to cut down the number and length of the tests that should be generated by means of selection criteria.

Test selection criteria can be functional, stochastic, structural, or fault-based [14]. Structural criteria exhibit the advantage that they allow for a fully automated generation of test cases. The downside is that, because they are not related to (realistic) fault models, their fault detection ability is usually restricted.

In order to capture the notion of possible fault, fault models can also be used for the derivation of tests [11]. The idea is to hypothesize the presence of a specific fault in a system and derive a test case that would find the respective fault *if it was present*. Fault models are used for the description of singular faults whose presence would be catastrophic, and also for the description of recurring faults. Typical examples include stuck-by-1 and division-by-zero. In a model-based context, approaches that operationalize fault models often inject faults and then use the mutated model to derive test cases that potentially violate given properties [3].

We have thus in principle two possible starting points: 1) because a model is incorrect, we have one or more counterexamples and want to check whether a SUT based on the specification is also flawed; or 2) we have a secure model and want to check if the SUT is secure as well. The first case is trivial: in case of one or more refinements, we already have two "interesting" abstract traces to start from. But what about the second case?

4.1   Unwinding-Based Coverage Criteria

Given a secure model, and without any further information, we need to assume that any pair of states in the EFSM (S_1, S_2) is potentially output inequivalent for low events in the implementation. However, we know that the 'interesting' pairs of states are precisely the output-inconsistent pairs in the Unwinding relation. To try to reach all interesting parts of the model when generating test-cases, we define the following coverage criteria (similarly as structural coverage criteria for state-charts [10]).

Coverage of pairs in the Unwinding relation: All pairs of states in the Unwinding relation R must be tested at least once, meaning a pair of low-equivalent but high-inequivalent input traces reaching S_1 and S_2 for $(S_1, S_2) \in R$ must be refined and tested; *Coverage of low inputs:* Given a pair $(S_1, S_2) \in R$ all low input symbols in $S_1 \cup S_2$ must be tested, meaning that for each symbol L that is accepted in S_1 or S_2 there is at least one pair of low-equivalent input traces reaching S_1 and S_2 such that the last input is L in both traces; *Loop coverage:* Given a pair $(S_1, S_2) \in R$ such that there exist paths with loops from the initial state to S_1 and S_2. Then pairs of low-equivalent traces of length up to k are refined and tested.

4.2 Non-interference and Mutation Testing

In order to generate interesting test cases, Büchler et al. [3] propose to mutate secure specifications by introducing faults that correspond to common implementation bugs with an impact to security properties. If M' is the result of such a mutation to a model M, then there are two possibilities. If $M' \models \varphi$, this M' is useless for test case generation. If, in contrast, $M' \not\models \varphi$, then the model checker will return counterexamples. If the SUT contains the fault, it will (likely) be found. If not, the test case is still valuable because it tests for a *potential* fault, in this case, a violation of the security property φ.

Mutation operators that break a secure unwinding relation can be easily constructed by violating the output consistency of related states. A full empirical investigation of this approach is challenging, since *any* output inconsistency will lead to a confidentiality/integrity violation. Identifying realistic mutation operators, that reflect common mistakes made by designers/programmers in this context, is the subject of ongoing work.

5 Preliminary Evaluation

We now compare a prototypical implementation of the unwinding-based model-checking approach against a naive (bounded) brute-forcing approach similar to self-composition, which yields preliminar evidence on the usefulness of our approach. The control algorithm considers all pairs of nodes (S_i, S_j) in the model and compute all symbolic traces reaching S_i and S_j respectively that are low equivalent. Then it solves the constraints (including output inconsistency) to force a counterexample.

We compare the execution time of our implementation (CTNI) of the presented algorithm against the brute-force approach for the discounted consumption example presented before, and a further example consisting of an access control scenario and its mutated insecure counterpart (Profile). Moreover we consider a bigger random model and its secure and insecure versions.

The results of the experiment are depicted in Fig. 2. We outperform brute-force as expected for *secure* models: because of the soundness of the Extended unwinding approach, it suffices to check the model statically without constraint

Model	Depth	CTNI	Brute-force
Discount secure	7	0.2s	N/A
Discount insecure	7	144s	192s
Profile secure	8	5ms	33.4s
Profile insecure	7	334s	679s
Random 5 states secure	8	1.9ms	45.3s
Random 5 states insecure	7	124s	182s

Fig. 2. Comparison of our approach against brute-forcing

solving, and the verification time stays constant regardless of the path depth. For the insecure versions, we perform similarly for small models, but we outperform the brute-force approach as the maximum path depth gets higher.

6 Related Work

Barthe et al. [1] proposed a characterization of non-interference that allows to express it in terms of a single program run. This is realized by self-composing a program with itself, and triggered applications also for the event-based, reactive information-flow domain. Recently, Clarkson et al. [4] proposed an extension to LTL, called HyperLTL, that allows to specify properties of sets of traces, including non-interference. They also report on the construction of a prototypical model-checker for a subset of their logic which uses off-the-shelve model-checkers and is based on the idea of model self-composition.

In [12] Roscoe et al. propose a characterization of non-interference in terms of determinism. They formalize these properties in CSP and claim that those can be efficiently verified by means of the FDR model-checker. However no complexity analysis or experimental data is available. Foccardi and Gorrieri [5] have also studied information flow in terms of process algebras. In particular, they generalize non-interference for non-deterministic systems and show how standard bisimulation checking algorithms can be used to check them. In [13] Ryan et al. discuss different extensions to non-interference for non-deterministic notions, including the afore mentioned approaches, and provide a proof that the bisimulation approaches correspond to unwinding theorems.

In [2] Bohannon et al. proposed a notion of reactive non-interference for programs and a type system to guarantee their security, based on a bisimulation property. This approach is comparable to the extended unwinding of [9], since the focus is on a sound approximation to non-interference. However, they do not provide a methodology to compute counterexamples.

To the best of our knowledge, there are no works in MBT considering information flow properties. In language-based security, an approach to randomly test assembly like programs has been proposed by Hritcu et. al. [8]. Different from their work we focus on a higher abstraction level, using coverage criteria and mutation.

7   Conclusions

In this paper we have presented a bounded model-checking approach based on an unwinding algorithm for EFSMs that allows one to compute counter-examples for non-interference. We also discuss how these results can be useful to define test-cases for exercising implementations. We have carried out a preliminary evaluation of our approach in terms of efficiency. As a result, we have observed that our approach performs considerably better than brute-force for secure models and as the maximum length increases.

There exist several directions towards extending this work: first, we plan to generalize our approach for weaker security notions such as declassification policies. On the other hand further applications of mutation-based testing to this setting are being considered. Finally, an implementation guiding a general purpose model-checker to take into account the unwinding analysis, would allow for a thorough comparison against self-composition based approaches.

Acknowledgements. This research was supported by the EU Project no. 257876 *SPaCIoS*: Secure Provision and Consumption in the Internet of Services.

References

1. Barthe, G., D'Argenio, P.R., Rezk, T.: Secure information flow by self-composition. In: CSF 2004, pp. 100–114. IEEE (2004)
2. Bohannon, A., Pierce, B.C., Sjöberg, V., Weirich, S., Zdancewic, S.: Reactive non-interference. In: CCS 2009, pp. 79–90. ACM (2009)
3. Buchler, M., Oudinet, J., Pretschner, A.: Spacite–web application testing engine. In: ICST 2012, pp. 858–859. IEEE (2012)
4. Clarkson, M.R., Finkbeiner, B., Koleini, M., Micinski, K.K., Rabe, M.N., Sánchez, C.: Temporal logics for hyperproperties. In: Abadi, M., Kremer, S. (eds.) POST 2014 (ETAPS 2014). LNCS, vol. 8414, pp. 265–284. Springer, Heidelberg (2014)
5. Focardi, R., Gorrieri, R.: The compositional security checker: A tool for the verification of information flow security properties. IEEE Transactions on Software Engineering 23(9), 550–571 (1997)
6. Goguen, J.A., Meseguer, J.: Security policies and security models. In: IEEE Symposium on Security and Privacy, pp. 11–20 (1982)
7. Goguen, J.A., Meseguer, J.: Unwinding and inference control. In: IEEE Symposium on Security and Privacy (1984)
8. Hritcu, C., Hughes, J., Pierce, B.C., Spector-Zabusky, A., Vytiniotis, D., de Amorim, A.A., Lampropoulos, L.: Testing noninterference, quickly. In: ICFP (2013)
9. Ochoa, M., Jürjens, J., Cuéllar, J.: Non-interference on UML State-charts. In: Furia, C.A., Nanz, S. (eds.) TOOLS 2012. LNCS, vol. 7304, pp. 219–235. Springer, Heidelberg (2012)
10. Offutt, J., Abdurazik, A.: Generating tests from UML specifications. In: France, R.B. (ed.) UML 1999. LNCS, vol. 1723, pp. 416–429. Springer, Heidelberg (1999)
11. Pretschner, A., Holling, D., Eschbach, R., Gemmar, M.: A generic fault model for quality assurance. In: Moreira, A., Schätz, B., Gray, J., Vallecillo, A., Clarke, P. (eds.) MODELS 2013. LNCS, vol. 8107, pp. 87–103. Springer, Heidelberg (2013)

12. Roscoe, A., Woodcock, J., Wulf, L.: Non-interference through determinism. In: Gollmann, D. (ed.) ESORICS 1994. LNCS, vol. 875, pp. 31–53. Springer, Heidelberg (1994)
13. Ryan, P.Y., Schneider, S.A.: Process algebra and non-interference. Journal of Computer Security 9(1), 75–103 (2001)
14. Utting, M., Pretschner, A., Legeard, B.: A taxonomy of model-based testing approaches. Softw. Test., Verif. Reliab. 22(5), 297–312 (2012)

Idea: State-Continuous Transfer of State in Protected-Module Architectures

Raoul Strackx and Niels Lambrigts

iMinds-DistriNet, University of Leuven, Leuven, Belgium
raoul.strackx@cs.kuleuven.be
niels.lambrigts@gmail.com

Abstract. The ability to copy data effortlessly poses significant security issues in many applications; It is difficult to safely lend out music or e-books, virtual credits cannot be transferred between peers without contacting a central server or co-operation with other network nodes, . . .

Protecting digital copies is hard because of the huge software and hardware trusted computing base applications have to rely on. Protected-module architectures (PMAs) provide an interesting alternative by relying only on a minimal set of security primitives. Recently it has been proven that such platforms can provide strong security guarantees. However, transferring state of protected modules has, to the best of our knowledge, not yet been studied.

In this paper, we present a protocol to transfer protected modules from one machine to another state-continuously; From a high level point of view, only a *single instance* of the module exists that executes without interruption when it is transferred from one machine to another. In practice however an attacker may (i) crash the system at any point in time (i.e., a crash attack), (ii) present the system with a stale state (i.e., a rollback attack), or (iii) trick both machines to continue execution of the module (i.e., a forking attack). We also discuss use cases of such a system that go well beyond digital rights management.

Keywords: Protected Module Architecture, State-Continuity, Transferring State.

1 Introduction

Computer science has transformed the world more rapidly than any technology before. Online encyclopedia enable worldwide access to knowledge, news travels faster than ever before and social media provide a global discussion platform. One key reason for this success is that data can be copied cheaply, easily and without loss of quality.

Unfortunately this ability also poses a security risk in many use cases. Digital rights management is the most obvious example: Once released, content providers cannot prevent that their data is distributed further. But many other use cases exist as well. We elaborate on valuable applications in Section 2.

F. Piessens et al. (Eds.): ESSoS 2015, LNCS 8978, pp. 43–50, 2015.

Certain types of data should resemble physical objects; It should not be possible to create exact copies and once transferred to another party, the sender should no longer have access. This is hard to guarantee in practice. Security checks can be added to applications and operating systems, but commodity operating systems are so complex that their correctness cannot be guaranteed. A determined attacker is likely to find an exploitable vulnerability (e.g, buffer overflows [13,22]) in this huge trusted computing base (TCB). Alternatively, the owner itself could have an incentive to break the implemented security features and launch physical attacks against the machine (e.g., cold boot attacks) [5,4].

Protected-Module Architectures. Recent advances in security architectures provide the required building blocks for an alternative approach. Protected-module architectures (PMAs) avoid a huge TCB by only providing a minimal set of security properties [12,6,14,11,8,21,20,24,10]. The exact set depends on the exact implementation, but all provide complete isolation of software modules. The PMA guarantees that modules have full control over their own memory regions; Any attempt to access memory locations belonging to the protected module at any privilege level (including from other modules), will be prevented. Protected modules can only be accessed through the interface that they expose explicitly.

Protected-module architectures can be used to harden security-sensitive parts of an application. Strackx et al. [20] evaluate a simple use case of a client connecting to a protected module. By placing SSL logic inside the protected module, only SSL packets cross the module's protection boundaries. Operating system services are still used to send and receive network packets, but these are *not* trusted. As any attempt to access sensitive memory locations in the protected module will be prevented by the PMA, this effectively reduces the power of a kernel-level attacker to that of a network-level attacker; Messages can be intercepted, modified, replayed or dropped, but sensitive data cannot be intercepted by an attacker.

Recently proposed protected-module architectures [14,6] also protect against sophisticated hardware attacks. Intel SGX, a PMA that is expected to be implemented in Intel processors in the near future, for example, guarantees that protected modules are only stored unencrypted when they reside in the CPU's cache. Before they are evicted to main memory or later to swap disk, they are confidentiality, integrity and version protected. This prevents many hardware attacks such as a cold boot attacks [5].

State-Continuity Guarantees. Agten et al. [2,1] and Patrignani et al. [17,16] formally proved that protected-module architectures can guarantee strong security properties of modules *while they execute continuously*. In practice however machines crash, need to reboot or lose power at unexpected times. To deal with such events, modules must store their state on disk. However, confidentiality and integrity protecting module states before they are passed to the untrusted operating system for storage, is not sufficient. After the system reboots modules cannot distinguish between fresh and stale states. In many applications this forms a security vulnerability.

Parno et al. [15] and Strackx et al. [18, 19] propose a solution and guarantee state-continuous execution; Modules either (eventually) advance with the provided input or never advance at all.

Our contributions. In this paper we build upon these security primitives and present a protocol to state-continuously transfer state of protected modules from one machine to another. At a high level, programmer's point of view, protected modules execute without interruption while they are transferred between machines. In practice however, we assume that an attacker may gain kernel- or hypervisor-level access to the system, but the underlying guarantees provided by the protected-module architecture cannot be broken. This implies that an attacker can (i) crash the system at any point in time, (ii) replay network messages, (iii) impersonate a remote party, and (iv) attempt to roll back the state of a module (e.g, by creating a new module and replay network packets). Since such a powerful attacker can easily launch denial-of-service attacks (e.g, by corrupting the kernel image), such attacks are not considered.

2   Use Cases

When it can be guaranteed that a protected-module instance cannot be rolled back nor forked when it is transferred to other machines, previously infeasible use cases become available.

Changing Machine. Parno et al. [15] and Strackx et al. [18] proposed an algorithm to guarantee state-continuous execution of protected modules. While this property enables versatile use cases, it prevents protected modules to be passed to another machine. In practice however, machines need to be replaced. Our protocol provides such support. Note that backups do *not* provide a good use case as they imply that a module could be restored to a previous state. We explicitly wish to defend against such behavior.

Multi Processor-Package Systems & Cloud Computing. Intel SGX provides strong security guarantees in face of software and hardware attackers by ensuring that protected modules are only stored unencrypted in the CPU's cache. When they are evicted to untrusted RAM or swap disk, they are confidentiality, integrity and version protected. Unfortunately, this also prevents protected modules to be transferred from one processor package to another on the same machine. A state-continuous enclave must always be executed on the same CPU package as it was created on. This poses significant practical challenges as applications can no longer be migrated from one CPU package to another to load balance execution load.

A similar problem occurs in a cloud computing setting where a virtual machine may be migrated from one physical server to another. State-continuous transfer of states can solve such problems.

Digital Wallets. Cryptocurrencies such as Bitcoin continue to gain traction. Bitcoins can be transferred from one user to another, but network consensus must be reached before the transfer can be asserted. This has the disadvantage that bitcoins cannot be transferred when the sender cannot connect to the network.

By relying on security properties of protected-module architectures, an alternative infrastructure can be built easily. Virtual credits can be stored as a simple counter in a protected module, completely isolated from the rest of the system. Transferring credits between peers can then be achieved using a state-continuous transfer of a portion of the available credits. The protocol presented in section 3 can be trivially modified to enable such use cases (i.e., the sender no longer sends its current state, nor will it permanently disable itself after the protocol completed).

Distributed Capability Systems. King-Lacroix et al. [7] propose BottleCap a capability-based mechanism for distributed networks. Resources can be accessed *iff* the user has the capability to do so. As access right checks are always local to the data that is accessed, such a system is much more scalable than traditional user-based authentication.

By placing capabilities in protected modules and transferring their state continuously, BottleCap's difficulties can be overcome; Capabilities can be easily revoked and special transfer policies for capabilities can be implemented easily.

Lending Digital Content. Using the state-continuity properties we provide, digital content can be lent similar to physical objects; Once content is passed, it can no longer be accessed by the sender. A book, for example, could be lent out by state-continuously transferring the protected module it is stored in. When the user wishes to read a page, it is rendered inside the protected module and only the resulting image is passed to unprotected code. While significantly raising the bar for attackers, we acknowledge that this setup does not prevent attackers from copying the book by copying each rendered page. Additional technologies such as Intel IPT[1] can be used to mitigate such attacks.

3   Transferring State

The protocol that we present only relies on properties (that can be) provided by almost all protected-module architectures. Therefore, we will not assume any protected-module architecture in particular when presenting our protocol.

Say we wish to transfer the state of a protected module M_{src} from one machine (i.e., src) to another (i.e., dst). Our protocol operates in two phases. In the first phase a new instance of module M with a "blank" state module is created on dst (called M_{dst}) and a public-private key pair is generated. When M_{src} is guaranteed that the module was deployed correctly, it commits to the state transfer. This marks the beginning of the second phase where the state is transferred to M_{dst}. By encrypting the state with M_{dst}'s public key, *only* M_{dst} is able to ever resume M's execution.

---

[1] http://ipt.intel.com/

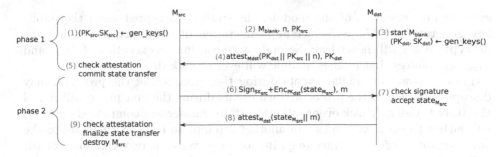

Fig. 1. Overview of the protocol where M_{src}'s state is transferred continuously to M_{dst}

The Protocol. Figure 1 displays the protocol graphically in more detail. In the first step, M_{src} generates a public-private key pair (PK_{src}, SK_{src}) and passes PK_{src} to dst together with M_{blank} (the module's code with a "blank" state) and a nonce n. Endpoint dst loads M_{blank} in memory and starts its execution with PK_{src} as argument. This public key will later be used to ensure that only states signed with the related private key will be accepted. The newly created module M_{dst} also generates a public-private key pair (PK_{dst}, SK_{dst}) that will be used to transfer M_{src}'s state securely. The three resulting keys are all stored in M_{dst} after which its correct execution is attested[2] to M_{src} (step 4). The enclosed nonce ensures freshness.

After M_{src} verified the attestation and found correct, it stops servicing user requests and uses the PMA's state-continuity property to commit to the state transfer; At some point in the future M_{src} will transfer it's state to a module M with possession of SK_{dst}, or the module will never advance it's state again. In step 6 M_{src}'s state is encrypted with PK_{dst} and signed with SK_{src}. It is passed together with a new nonce m to dst. After ensuring that the passed state originated from M_{src}, M_{dst} decrypts the state and resumes the module's execution. To avoid that M_{src} continues to attempt to transfer state, the protocols terminates by attesting that M_{dst} accepted the state transfer. M_{src} can now permanently disable requests to retry state transfer to M_{dst} and destruct itself.

Security Analysis. Safety of our protocol is based on two simple properties. First, we rely on the ability to create and store cryptographic keys in protected modules. At the beginning of phase 2 (i.e., after step 5), M_{src} is committed to transfer state from M_{src} to M_{dst}. To ensure that only one module instance can

---

[2] An attestation is a log of inputs and outputs that is signed by a trusted entity (e.g., a PMA platform or TPM chip). In a simple implementation a private key is embedded in the trusted entity that is signed by a trusted third party. This allows easy verification of the attestation log. Implementations such as Intel SGX [3] however use more complex cryptographic protocols to prevent linkability of attestation logs. We do not target a single PMA specifically in this section but simply state what needs to be attested.

ever resume execution of the module, the state is encrypted using the public key generated by M_{dst}. Correct implementation of M_{dst} guarantees that the decryption key will never leak. Second, state-continuous execution of M_{src} and M_{dst} guarantees that neither module will ever be forked.

Loss of power or similar events during the execution of the protocol, may disrupt execution of the protocol. Such events during the first phase will cancel the state transfer. Attack events after the state transfer was committed however, will either (i) eventually result in another attempt to transfer the state to the *same remote module instance* (e.g., in case of power or network failure), or (ii) will lead to a situation where the module will never resume execution (e.g., machine dst is physically destroyed).

To ensure that the state originated from M_{src}, the protocol passes in step 2 M_{src}'s public key to M_{dst}. This additional security check is not strictly required, but omitting it leaves the protocol open to a denial-of-service attack from a remote attacker. As M_{dst} then would then accept *any* provided state (and only accepts a new state once), a remote attacker could send it any state. In that case M_{src} would never be able to complete its state transfer, but it also does not allow user requests to be handled until the state is transferred.

While guaranteeing that state can be transferred securely, the protocol does *not* provide any assurance about the origin of the module's state. An attacker masquerading as M_{src} could even fabricate a state. In most cases this does not pose a vulnerability as it only needs to be guaranteed that a specific module instance will continue to execute continuously while it is transferred from host to host. For other use cases, a proof of the origin of the module can be easily added in two ways: (1) Verification of the origin of a module could be built in the module itself. When the module is initially created, for example, it could generate a secret shared with a verifier. Even when the module is transferred from host to host, possession of this secret proves its origin. Or (2) the protocol could be modified to also attest the correct set up of M_{src} to M_{dst} in step 2. This would prevent the fabrication of states by an attacker, but a verifier interacting with two module's over time cannot determine whether they are the same module instance.

4   Related Work

State-continuous transfer of state is easy when the user can rely on the correctness of the operating system. In practice this assumption is hard to guarantee. Protected-module architectures provide an interesting alternative, but to the best of our knowledge, state-continuous transfer has not been addressed in such a setting.

Related work that relies on a very limited TCB exist, but only provide very specific security guarantees. Van Dijk et al. [23], for example, present a system where a central server provide tamper-evident persistent storage. Rollback and forking attacks of this data is prevented using a trusted time stamping device executing on an untrusted third party's server. Our protocol can achieve similar guarantees (i.e., clients can pass data with build-in protection against rollback or forking) but is more flexible. Modules for example, can easily choose to only transfer their state partially, as was discussed in the digital wallet example of Section 2. We also do not rely on a central server, which provides significant scalability advantages.

More recently Kotla [9] presented a way to build the digital equivalent of scratch-off cards. A client can choose to either use a cryptographic key stored in the TPM chip but cannot hide its use upon an audit. Or it does not access the cryptographic key and after this choice is proven to a verifier, access is permanently revoked. Such scatch-off cards can, for example, be used to download digital media and later request a refund if it was never accessed. Again, our protocol supports similar, but more general use cases; Data can be accessed until it is transferred to another machine.

5 Conclusion

The ability to endlessly copy data poses security challenges in many settings. We presented a protocol to state-continuously transfer state of protected modules from one machine to another. We believe that this new security feature of protected-module architectures enables many use cases that were not possible before.

References

1. Agten, P., Jacobs, B., Piessens, F.: Sound modular verification of c code executing in an unverified context. Accepted for publication in Proceedings of the 42nd Annual ACM SIGPLAN-SIGACT Symposium on Principles of Programming Languages (POPL 2015) (January 2015)
2. Agten, P., Strackx, R., Jacobs, B., Piessens, F.: Secure compilation to modern processors. In: 2012 IEEE 25th Computer Security Foundations Symposium (CSF 2012), pp. 171–185. IEEE Computer Society, Los Alamitos (2012)
3. Anati, I., Gueron, S., Johnson, S., Scarlata, V.: Innovative technology for CPU based attestation and sealing. In: Proceedings of the 2nd International Workshop on Hardware and Architectural Support for Security and Privacy (2013)
4. Chan, E.M., Carlyle, J.C., David, F.M., Farivar, R., Campbell, R.H.: BootJacker: Compromising computers using forced restarts. In: Proceedings of the 15th ACM Conference on Computer and Communications Security, CCS 2008, pp. 555–564. ACM, New York (2008)
5. Halderman, J., Schoen, S., Heninger, N., Clarkson, W., Paul, W., Calandrino, J., Feldman, A., Appelbaum, J., Felten, E.: Lest we remember: Cold boot attacks on encryption keys. In: USENIX Security Symposium, pp. 45–60 (2008)
6. Intel Corporation. Software Guard Extensions Programming Reference (2013)
7. King-Lacroix, J., Martin, A.: Bottlecap: A credential manager for capability systems. In: Proceedings of the Seventh ACM Workshop on Scalable Trusted Computing, STC 2012, pp. 45–54. ACM, New York (2012)

8. Koeberl, P., Schulz, S., Sadeghi, A.-R., Varadharajan, V.: Trustlite: a security architecture for tiny embedded devices. In: Proceedings of the Ninth European Conference on Computer Systems (EuroSys 2014), p. 10. ACM (2014)
9. Kotla, R., Rodeheffer, T., Roy, I., Stuedi, P., Wester, B.: Pasture: secure offline data access using commodity trusted hardware. In: Proceedings of the 10th USENIX Conference on Operating Systems Design and Implementation (OSDI 2012) (2012)
10. McCune, J.M., Li, Y., Qu, N., Zhou, Z., Datta, A., Gligor, V., Perrig, A.: TrustVisor: Efficient TCB reduction and attestation. In: Proceedings of the IEEE Symposium on Security and Privacy (S&P 2010) (May 2010)
11. McCune, J.M., Parno, B., Perrig, A., Reiter, M.K., Isozaki, H.: Flicker: An execution infrastructure for TCB minimization. In: Proceedings of the ACM European Conference in Computer Systems (EuroSys), pp. 315–328. ACM (April 2008)
12. Noorman, J., Agten, P., Daniels, W., Strackx, R., Herrewege, A.V., Huygens, C., Preneel, B., Verbauwhede, I., Piessens, F.: Sancus: Low-cost trustworthy extensible networked devices with a zero-software trusted computing base. In: 22nd USENIX Security Symposium (Usenix 2013). USENIX Association (August 2013)
13. One, A.: Smashing the stack for fun and profit. Phrack Magazine 7(49) (1996)
14. Owusu, E., Guajardo, J., McCune, J., Newsome, J., Perrig, A., Vasudevan, A.: OASIS: on achieving a sanctuary for integrity and secrecy on untrusted platforms. In: Conference on Computer & Communications Security (CCS 2013) (2013)
15. Parno, B., Lorch, J.R., Douceur, J.R., Mickens, J., McCune, J.M.: Memoir: Practical state continuity for protected modules. In: Proceedings of the IEEE Symposium on Security and Privacy (S&P 2011) (May 2011)
16. Patrignani, M., Agten, P., Strackx, R., Jacobs, B., Clarke, D., Piessens, F.: Secure compilation to protected module architectures. Accepted for publication in Transactions on Programming Languages and Systems, TOPLAS (2014)
17. Patrignani, M., Clarke, D., Piessens, F.: Secure Compilation of Object-Oriented Components to Protected Module Architectures. In: Shan, C.-c. (ed.) APLAS 2013. LNCS, vol. 8301, pp. 176–191. Springer, Heidelberg (2013)
18. Strackx, R., Jacobs, B., Piessens, F.: ICE: A passive, high-speed, state-continuity scheme. In: Annual Computer Security Applications Conference (ACSAC 2014) (2014)
19. Strackx, R., Jacobs, B., Piessens, F.: ICE: A passive, high-speed, state-continuity scheme (extended version). CW Reports CW672, KU Leuven (August 2014)
20. Strackx, R., Piessens, F.: Fides: Selectively hardening software application components against kernel-level or process-level malware. In: Computer and Communications Security (CCS 2012) (October 2012)
21. Strackx, R., Piessens, F., Preneel, B.: Efficient Isolation of Trusted Subsystems in Embedded Systems. In: Jajodia, S., Zhou, J. (eds.) SecureComm 2010. LNICST, vol. 50, pp. 344–361. Springer, Heidelberg (2010)
22. Strackx, R., Younan, Y., Philippaerts, P., Piessens, F., Lachmund, S., Walter, T.: Breaking the memory secrecy assumption. In: Proceedings of the Second European Workshop on System Security, pp. 1–8. ACM (2009)
23. van Dijk, M., Rhodes, J., Sarmenta, L.F.G., Devadas, S.: Offline untrusted storage with immediate detection of forking and replay attacks. In: Proceedings of the 2007 ACM Workshop on Scalable Trusted Computing, STC 2007 (2007)
24. Vasudevan, A., Chaki, S., Jia, L., McCune, J., Newsome, J., Datta, A.: Design, implementation and verification of an extensible and modular hypervisor framework. In: Proceedings of the 2013 IEEE Symposium on Security and Privacy, SP 2013, pp. 430–444. IEEE Computer Society, Washington, DC (2013)

Are Your Training Datasets Yet Relevant?

An Investigation into the Importance of Timeline in Machine Learning-Based Malware Detection

Kevin Allix, Tegawendé F. Bissyandé, Jacques Klein, and Yves Le Traon

SnT - University of Luxembourg

Abstract. In this paper, we consider the *relevance of timeline* in the construction of datasets, to highlight its impact on the performance of a machine learning-based malware detection scheme. Typically, we show that simply picking a random set of known malware to train a malware detector, as it is done in many assessment scenarios from the literature, yields *significantly biased* results. In the process of assessing the extent of this impact through various experiments, we were also able to confirm a number of intuitive assumptions about Android malware. For instance, we discuss the existence of Android malware lineages and how they could impact the performance of malware detection in the wild.

1 Introduction

Malware detection is a challenging endeavor in mobile computing, where thousands of applications are uploaded everyday on application markets [1] and often made available for free to end-users. Market maintainers then require efficient techniques and tools to continuously analyze, detect and triage malicious applications in order to keep the market as clean as possible and maintain user confidence. For example, Google has put in place a number of tools and processes in the Google Play official market for Android applications. However, using antivirus software on large datasets from Google reveals that hundreds of suspicious apps are still distributed incognito through this market [2].

Unfortunately, malware pose various threats that cannot be ignored by users, developers and retailers. These threats range from simple user tracking and leakage of personal information [3], to unwarranted premium-rate subscription of SMS services, advanced fraud, and even damaging participation to botnets [4]. To address such threats, researchers and practitioners increasingly turn to new techniques that have been assessed in the literature for malware detection in the wild. Research work have indeed yielded promising approaches for malware detection. A comprehensive survey of various techniques can be found in [5]. Approaches for large-scale detection are often based on Machine learning techniques, which allow to sift through large sets of applications to detect anomalies based on measures of similarity of features [6,7,8,9,10,11,12,13,14].

To assess malware detection in the wild, the literature resorts to the 10-Fold Cross validation scheme with datasets that we claim are biased and yield biased results. Indeed, various aspects of construction of training datasets are usually overlooked. Among such aspects is the *history aspect* which assumes that the

F. Piessens et al. (Eds.): ESSoS 2015, LNCS 8978, pp. 51–67, 2015.

training dataset, which is used for building classifiers, and the test dataset, which is used to assess the performance of the technique, should be *historically coherent*: the former must be historically anterior to the latter. This aspect is indeed a highly relevant constraint for real-world use cases and we feel that evaluation and practical use of state-of-the-art malware detection approaches must follow a process that mimics the history of creation/arrival of applications in markets as well as the history of appearance of malware: *detecting malware before they are publicly distributed in markets is probably more useful than identifying them several months after they have been made available.*

Nevertheless, in the state-of-the-art literature, the datasets of evaluation are borrowed from well-known labelled repositories of apps, such as the Genome project, or constructed randomly, using market-downloaded apps, with the help of Antivirus products. However, the history of creation of the various apps that form the datasets are rarely, if ever, considered, leading to situations where **some items in the training datasets are** *"from the future"*, **i.e., posterior, in the timeline, to items in the tested dataset**. Thus, different research questions are systematically eluded in the discussion of malware detector performance:

RQ-1. Is a randomly sampled training dataset equivalent to a dataset that is historically coherent to the test dataset?

RQ-2. What is the impact of using malware knowledge "from the future" to detect malware in the present?

RQ-3. How can the potential existence of families of malware impact the features that are considered by machine learning classifiers?

RQ-4. How *fresh* must be the apps from the training dataset to yield the best classification results?

RQ-5. Is it sound/wise to account for all known malware to build a training dataset?

This Paper. We propose in this paper to investigate the effect of ignoring/ considering historical coherence in the selection of training and test datasets for malware detection processes that are built on top of Machine learning techniques. Indeed we note from literature reviews that most authors do not take this into account. Our ultimate aim is thus to provide insights for building approaches that are consistent with the practice of application –including malware– development and registration into markets. To this end, we have devised several typical machine learning classifiers and built a set of features which are textual representations of basic blocks extracted from the Control-Flow Graph of applications' byte-code. Our experiments are also based on a sizeable dataset of about 200,000 Android applications collected from sources that are used by authors of contributions on machine learning-based malware detection.

The contributions of this paper are:

- We propose a thorough study of the history aspect in the selection of training datasets. Our discussions highlight different biases that may be introduced if this aspect is ignored or misused.

- Through extensive experiments with tens of thousands of Android apps, we show the variations that the choice of datasets age can have on the malware detection output. To the best of our knowledge, we are the first to raise this issue and to evaluate its importance in practice.
- We confirm, or show how our experiments support, various intuitions on Android malware, including the existence of so-called lineages.
- Finally, based on our findings, we discuss (1) the assessment protocols of machine learning-based malware detection techniques, and (2) the design of datasets for training real-world malware detectors.

The remainder of this paper is organized as follows. Section 2 provides some background on machine learning-based malware detection and highlights the associated assumptions on dataset constructions. We also briefly describe our own example of machine-learning based malware detection. Section 3 presents related work to support the ground for our work. Section 4 describes the experiments that we have carried out to answer the research questions, and presents the take-home messages derived from our empirical study. We propose a final discussion on our findings in Section 5 and conclude in Section 6.

2 Preliminaries

The Android mobile platform has now become the most popular platform with estimated hundreds of thousands of apps in the official Google Play market alone and downloads in excess of billions. Unfortunately, as this popularity has been growing, so is malicious software, i.e., malware, targeting this platform. Studies have shown that, on average, Android malware remain unnoticed up to 3 months before a security researcher stumbles on it [15], leaving users vulnerable in the mean time. Security researchers are constantly working to propose new malware detection techniques, including machine learning-based approaches, to reduce this 3-months gap.

Machine Learning: Features & Algorithms: As summarized by Alpaydin, "Machine Learning is programming computers to optimize a performance criterion using example data or past experience" [16]. A common method of learning is known as *supervised* learning, a scheme where the computer is helped through a first step of *training*. The training data consists of Feature Vectors, each associated with a label, e.g., in our case, apps that are already known to be malicious (*malware* class) or benign (*goodware* class). After a run of the learning algorithm, the output is compared to the target output and learning parameters may be corrected according to the magnitude of the error. Consequently, to perform a learning that will allow a *classification* of apps into the malware and goodware classes, the approach must define a correlation measure and a discriminative function. The literature of Android malware detection includes diverse examples of features, such as n-grams of bytecode, API usages, application permission uses, etc. There also exist a variety of classification algorithms, including Support Vector Machine (SVM) [17], the RandomForest ensemble decision-trees algorithm [18], the RIPPER rule-learning algorithm [19] and the tree-based *C4.5* algorithm [20]. In our work, because we focus exclusively on the history aspect, we constrain all aforementioned variables to values that are widely used in the literature, or based on our own experiments which have allowed us to select the most appropriate settings. Furthermore, it is noteworthy that we do not aim for absolute performance, but rather measure performance delta between several approaches of constructing training datasets.

Working Example: We now provide details on the machine-learning approach that will be used as a working example to investigate the importance of history in the selection of training and test datasets. Practically, to obtain the features for our machine-learning processes, we perform static analysis of Android applications' bytecode to extract an abstract representation of the program's control-flow graph (CFG). We obtain a CFG that is expressed as character strings using a method devised by Pouik *et al.* in their work on establishing similarity between Android applications [21], and that is based on a grammar proposed by Cesare and Xiang [22]. The string representation of a CFG is an abstraction of the application's code; it retains information about the *structure* of the code, but discards low-level details such as variable names or register numbers. This property is desirable in the context of malware detection as two variants of a malware may share the same abstract CFG while having different bytecode. Given an application's abstract CFG, we collect all basic blocks that compose it and refer to them as the features of the application. A basic block is a sequence of instructions in the CFG with only one entry point and one exit point. It thus represents the smallest piece of the program that is always executed altogether. By learning from the training dataset, it is possible to expose, if any, the basic blocks that appear statistically more in malware.

The basic block representation used in our approach is a high-level abstraction of the atomic parts of an Android application. A more complete description of this feature set can be found in [23]. For reproducibility purposes, and to allow the research community to build on our experience, the data we used (full feature matrix and labels) is available on request.

Methodology: This study is carried out as a large scale experiment that aims at investigating the extent of the relevance of history in assessing machine learning-based malware detection. This study is important for paving the road to a true success story of trending approaches to Android malware detection. To this end, our work must rely on an extensive dataset that is representative of real-world Android apps and of datasets used in the state-of-the-art literature.

Dataset: To perform this study we collect a large dataset of android apps from various markets: $78,460$ (38.04%) apps from Google Play, $72,093$ (34.96%) from appchina, and $55,685$ (27.00%) from Other markets[1]. A large majority of our dataset comes from Google Play, the official market, and appchina.

An Android application is distributed as an .apk file which is actually a ZIP archive containing all the resources an application needs to run, such as the application binary code and images. An interesting side-effect of this package format

---

[1] *Other markets* include anzhi, 1mobile, fdroid, genome, etc.

Table 1. A selection of Android malware detection approaches

Approach	Year	Sources	Historical Coherence
DREBIN [6]	2014	"Genome, Google Play, Chinese and russian markets, VirusTotal	No
[24]	2013	"common Android Markets" for goodware, "public databases of antivirus companies" for malware	No
[13]	2012	Undisclosed	No
DROIDMAT [25]	2012	Contagio mobile for malware, Google Play for goodware	No
[26]	2013	Genome, VirusTotal, Google Play	No
[27]	2013	Contagio mobile and Genome for malware, Undisclosed for goodware	No
[28]	2013	"from official and third party Android markets" for Goodware, Genome for malware	No
[29]	2013	Google Play (labels from 10 commercial Anti virus scanners)	No

is that all the files that makes an application go from the developer's computer to end-users' devices without any modification. In particular, all metadata of the files contained in the .apk package, such as the last modification date, are preserved. All bytecode, representing the application binary code, is assembled into a *classes.dex* file that is produced at packaging-time. Thus the last modification date of this file represents the packaging time. In the remainder of this paper, packaging date and compilation date will refer to this date.

To infer the historical distribution of the dataset applications, we rely on compilation date at which the Dalvik[2] bytecode (*classes.dex* file) was produced. We then sent all the app packages to be scanned by virus scanners hosted by VirusTotal [3] . VirusTotal is a web portal which hosts about 40 products from renown anti virus vendors, including McAfee®, Symantec® or Avast®. In this study, an application is labelled as malware if at least one scanner flags it as such.

Machine learning Parameters: In all our experiments, we have used the parameters that provided the best results in a previous large-scale study [23]. Thus, we fixed the number of features to 5,000 and selected the 5,000 features with highest Information Gain values as measured on the training sets. The RandomForest algorithm, as implemented in the Weka[4] Framework, was used for all our experiments.

3   Related Work

In this section, we propose to revisit related work to highlight the importance of our contributions in this paper. We briefly present previous empirical studies and their significance for the malware detection field. Then we go over the literature of malware detection to discuss the assessment protocols.

Empirical studies: Empirical studies have seen a growing interest over the years in the field of computer science. The weight of empirical findings indeed help ensure that research directions and results are in line with practices. This is especially important when assessing the performance of a research approach. A large body of the literature has resorted to extensive empirical studies for devising a reliable experimental protocol [30, 31, 32]. Recently, Allix *et al.* have proposed a large-scale empirical studies on the dataset sizes used in the assessment

---

[2] Dalvik is the virtual machine running Android apps.
[3] https://www.virustotal.com
[4] http://www.cs.waikato.ac.nz/ml/weka/

of machine learning-based malware detection approaches [23]. In their work, the authors already questioned the assessment protocols used in the state-of-the-art literature. Guidelines for conducting sound Malware Detection experiments were proposed by Rossow *et al* [33]. Our work follows the same objectives, aiming to highlight the importance of building a reliable assessment protocol for research approaches, in order to make them more useful for real-world problems.

In the field of computer security, empirical studies present distinct challenges including the scarcity of data about cybercrimes. We refer the reader to a report by Böhme and Moore [34]. Recently, Visaggio *et al.* empirically assessed different methods used in the literature for detecting obfuscated code [35]. Our work is in the same spirit as theirs, since we also compare different methods of selecting training datasets and draw insights for the research community.

With regards to state-of-the-art literature tackled in this work, a significant number of Machine Learning approaches for malware detection [29,6,36,37,38,39] have been presented to the research community. The feature set that we use in this paper was evaluated in [23] and achieved better performance than those approaches. Thus, our experiments are based on a sound feature set for malware detection. We further note that in the assessment protocol of all these state-of-the-art approaches, the history aspect was eluded when selecting training sets.

Malware Detection & Assessments: We now review the assessment of malware detection techniques that are based on machine learning. For comparing performances with our own approach, we focus only on techniques that have been applied to the Android ecosystem. In Table 1, we list recent "successful" approaches from the literature of malware detection, and describe the origin of the dataset used for the assessment of each approach. For many of them, the applications are borrowed from known collections of malware samples or from markets such as Google Play. They also often use scanners from VirusTotal to construct the ground truth. In our approach, we have obtained our datasets in the same ways. Unfortunately, to the best of our knowledge and according to their protocol descriptions from the literature, none of the authors has considered clearly ordering the data to take into account the history aspect. It is therefore unfortunate that the high performances recorded by these approaches may never affect the fight against malware in markets.

In the remainder of this section we list significant related work examples, provide details on the size of their dataset and compare them to our history-unaware 10-Fold experiments. None of them has indeed taken into account the history aspect in their assessment protocol. In 2012, Sahs & Khan [13] built an Android malware detector with features based on a combination of Android-specific permissions and a Control-Flow Graph representation. Their classifier was tested with k-Fold [5] cross validation on a dataset of 91 malware and 2 081 goodware. Using permissions and API calls as features, Wu et al. [25] performed their experiments on a dataset of 1 500 goodware and 238 malware. In 2013, Amos et al. [26] leveraged dynamic application profiling in their malware detector. Demme et al. [27] also used dynamic application analysis to perform

---

[5] The value of k used by Sahs & Khan was not disclosed.

malware detection with a dataset of 210 goodware and 503 malware. Yerima et al. [28] built malware classifiers based on API calls, external program execution and permissions. Their dataset consists of 1 000 goodware and 1 000 malware. Canfora et al. [24] experimented feature sets based on SysCalls and permissions.

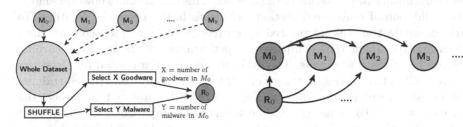

Fig. 1. Process of constructing a random training dataset R_0 for comparison with the training dataset constituted of all data from month M_0

Fig. 2. Classification process: the training dataset is either the dataset of a given month (e.g., M_0) or a random dataset constructing as in Figure 1

4 Experimental Findings

In this section, we report on the experiments that we have conducted, and highlight the findings. First we discuss to what extent it is important that datasets remain historically coherent, as opposed to being selected at random (cf. Section 4.1). This discussion is based on qualitative aspects as well as quantitative evaluation. Second, we conduct experiments that attempt to provide a hint to the existence of lineages in Android malware in Section 4.2. Subsequently, we investigate in Section 4.3 the bias in training with new data for testing with old data, and inversely. Finally, we investigate the limitations of a naive approach which would consist in accumulating information on malware samples as time goes, in the hope of being more inclusive in the detection of malware in the future (cf. Section 4.4).

4.1 History-Aware Construction of Datasets

As described in Section 2, a key step of machine-learning approaches is the training of classifiers. The construction of the corresponding training dataset is consequently of importance, yet details about how it is achieved are largely missing from the literature, as was shown in Section 3.

There are two common selection patterns for training datasets: (1) use a collected and published dataset of malware, such as Genome, to which one adds a subset of confirmed goodware; (2) build the dataset by randomly picking a subset of goodware and malware from a dataset collected from either an online market or an open repository. Both patterns lead to the same situations: i.e. that *some items in the training dataset may be historically posterior to items in the tested dataset.* In other words, (1) the construction of the training set is equivalent to a random history-unaware selection from a mix of known malware and goodware; and (2) the history of creation/apparition of android applications is not considered as a parameter in assessment experiments, although the practice of malware detection will face this constraint.

Following industry practices, when a newly uploaded set of applications must be analyzed for malware identification, the training datasets that are used are, necessarily, historically anterior to the new set. This constraint is however eluded in the validation of malware detection techniques in the research literature. To clearly highlight the bias introduced by current assessment protocols, we have devised an experiment that compares the performance of the machine learning detectors in different scenarios. The malware detectors are based on classifiers that are built in two distinct settings: either with randomly-constructed training datasets using a process described in Figure 1 or with datasets that respect the history constraint. To reduce the bias between these comparisons, we ensure that the datasets are of identical sizes and with the same class imbalance between goodware and malware. Thus to build a history-unaware dataset R_0 for comparing with training dataset constituted of data from month M_0, we randomly pick within the whole dataset the same numbers of goodware and malware as in M_0. We perform the experiments by training first on M_0 and testing on all following months, then by training on R_0 and testing on all months (cf. Figure 2).

Figure 3 illustrates the results of our experiments. When we randomly select the training dataset from the entire dataset and build classifiers for testing applications regrouped by month, the precision and recall values of the malware detector range between 0.5 and 0.85. The obtained F-Measure is also relatively high and roughly stable. This performance is in line with the performances of state-of-the-art approaches reported in the literature.

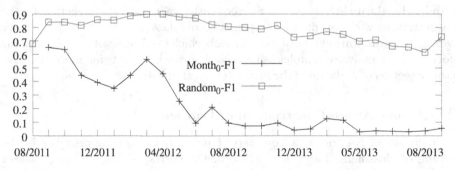

Reading: The $Month_0$ curve shows the F-Measure for a classfier trained on the month 0, while the $Random_0$ curve presents the F-Measure for a classifier built with a training set of same size and same goodware/malware ratio as month 0, but drawn randomly from the whole dataset.

Fig. 3. Performance of malware detectors with history-aware and with history-unaware selection of training datasets

We then proceed to constrain the training dataset to be historically coherent to the test dataset. We select malware and benign apps in the set of apps from a given month, e.g., M_0, as the source of data for building the training dataset for the classification. The tests sets remain the same as in the previous experiments, i.e., the datasets of applications regrouped by month. We observe that as we move away from M_0 to select test data, the performance considerably drops.

We have repeated this experiment, alternatively selecting each different month from our time-line as the month from which we draw the training dataset. Using a training set that is not historically coherent always led to significantly higher performance than using a historically coherent training set.

Finding RQ-1: *Constructing a training dataset that is consistent with the history of apparition of applications yields performances that are significantly worst than what is obtained when simply randomly collecting applications in markets and repositories. Thus,* **without further assessment, state-of-the-art approaches cannot be said to be powerful in real-world settings**.

Finding RQ-2: *With random selections, we allow malware "from the future" to be part of the training sets. This however leads to biased results since the performance metrics are artificially improved.*

4.2 Lineages in Android Malware

Our second round of experiments has consisted in investigating the capabilities of a training dataset to help build classifiers that will remain performant over time. In this step of the study we aim at discovering how the variety of malware is distributed across time. To this end, we consider building training datasets with applications in each month and test the yielded classifiers with the data of each following months.

Figures 4 and 5 provide graphs of the evolution over time of, on the one hand, F-Measure and, on the other hand, Precision of malware detectors that have been built with a training dataset at month M_i and applied on months $M_{k,k>i}$. Disregarding outliers which lead to the numerous abrupt rise and breaks in the curves, the yielded classifiers have, on average, a stable and high Precision, with values around 0.8. This finding suggests that *whatever the combination of training and test dataset months, the built classifiers still allow to identify with good precision the malware whose features have been learnt during training*.

On the other hand, the F-measure performance degrades over time: for a given month M_i whose applications have been used for the training datasets, the obtained classifier is less and less performant in identifying malware in the following months $M_{k,k>i}$. This finding, correlated to the previous one, suggests that, over time, the features that are learnt in the training dataset correspond less and less to all malware when we are in the presence of **lineages** in the Android malware. We define a `lineage` as a set of malware that share the same traits, whether in terms of behavior or of coding attributes. Note that we differentiate the term `lineage` from the term `family` which, in the literature, concern a set of malware that exploit the same vulnerability. *Lineage* is a more general term.

The experiments also highlight the bias introduced when training classifiers with a specific and un-renewed set of malware, such as the Genome dataset, which is widely used. It also confirms why the random selection of malware in the entire time-line as presented in Section 4.1, provides good performances:

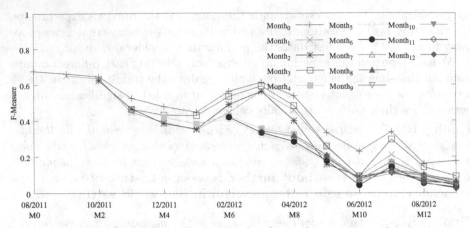

Fig. 4. Performance Evolution of malware detectors over time

many lineages are indeed represented in such training datasets, including lineages
that should have appeared for the first time in the test dataset.

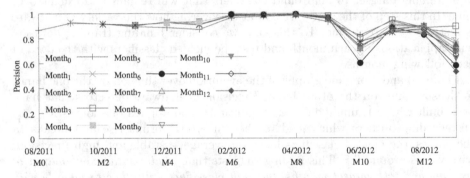

Fig. 5. Evolution of Precision of malware detectors over time

Finding-RQ3: *Android malware is diversified. The existence of lineages com-*
plicates malware detection, since training datasets must be regularly updated to
include a larger variety of malware lineages representatives.

4.3   Is Knowledge "from the future" the Grail?

Previous experiments have shown that using applications from the entire time-line,
without any historical constraint, favorably impacts the performance of malware
detectors. We have then proceeded to show that, when the training dataset is too
old compared to the test dataset, this performance drops significantly. We now investigate whether training data that are strictly posterior to the test dataset could
yield better performance than using data that are historically anterior (coherent).
Such a biased construction of datasets is not fair when the objective is to actively

keep malicious apps from reaching the public domain. However, such a construction can be justified by the assumption that the present might always contain representatives of malware lineages that have appeared in the past.

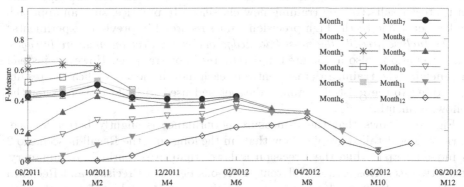

Fig. 6. Performance of malware detectors when using recent data to test on old datasets

In the Android ecosystem, thousands of applications are created weekly by developers. Most of them, including malware from new lineages, cannot be thoroughly checked. Nevertheless, after some time, antivirus vendors may identify the new malware. Machine-learning processes can thus be used to automate a large-scale identification of malware in applications that have been made available for some time. Figure 6 depicts the F-Measure performance evolution of the malware detectors: for each month M_i, that is used for training, the obtained classifiers are used to predict malware in the previous months $M_{k,k<i}$. Overall, the performance is dropping significantly with the time difference between test and training datasets.

Finding-RQ4: *Apps, including malware, used for training in machine learning-based malware detection must be historically close to the target dataset that is tested. Older training datasets cannot account for all malware lineages, and newer datasets do not contain enough representatives of malware from the past.*

4.4 Naive Approaches to the Construction of Training Datasets

Given the findings of our study presented in previous sections, we investigate through extensive experiments the design of a potential research approach for malware detection which will be in line with the constraints of industry practices. At a given time t, one can only build classifiers using datasets that are anterior to t. Nevertheless, to improve our chances of maintaining performance, two protocols can be followed:

(1) *Keep renewing the training dataset entirely to stay historically close to the target dataset of test. This renewal process must however be automated to remain realistic*: In this scenario, we assume that a bootstrap step is achieved with antivirus products at month M_0 to provide a first reliable training dataset.

The malware detection system is then on its own for the following months. Thus, the classification that is obtained on month M_1, using month M_0 for training, will be used "as is" to train the classifiers for testing applications data of month M_2. This system is iterated until month M_n as depicted in Figure 7, meaning that, once it is bootstrapped, the detection system is automated and only relies on its test results to keep training new classifiers. In practice, such an approach makes sense due to the high precision values recorded in previous experiments.

(2) *Include greedily the most knowledge one can collect on malware lineages*: This scenario is also automated and requires bootstrapping. However, instead of renewing the training dataset entirely each month, new classification results are added to the existing training dataset and used to build classifiers for the following month.

Figure 8 shows that the F-measure performance is slightly better for scenario 2. The detailed graphs show that, in the long run, the Recall in scenario 2 is indeed better while the Precision is lower than in scenario 1. In summary, these two scenarios exhibit different trade-offs between Precision and Recall in the long run: Scenario 1 manages to pinpoint a small number of malware with good precision while scenario 2 instead finds more malware at the cost of a higher false-positive rate.

While of little use in isolation, those scenarios provide insights through empirical evidence on how machine learning-based malware detection systems should consider the construction of training sets.

Finding-RQ5: *Maintaining performance of malware detectors cannot be achieved by simply adding/renewing information in training datasets based on the output of previous runs. However, these scenarios have shown interesting impact on performance evolution over time, and must be further investigated to identify the right balance.*

Fig. 7. Using classification results of M_{n-1} as training dataset for testing M_n

5   Insights and Future work

Findings. (1) History constraints must not be eluded in the construction of training datasets of machine learning-based malware detectors. Indeed, they introduce significant bias in the interpretation of the performance of malware classifiers. (2) There is a need for building a reliable, and continuously updated, benchmark for machine learning-based malware detection approaches. We make available, upon request, the version we have built for this work. Our benchmark dataset contains about 200,000 Android applications spanning 2 years of historical data of Android malware.

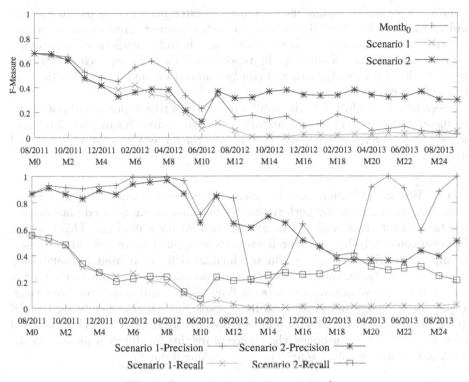

Fig. 8. Comparing two naive approaches

Insights. (1) Machine-learning cannot assure the identification of an entirely new lineage of malware that is not represented in the training dataset. Thus, there is need to regularly seed the process with outside information, such as from antivirus vendors, of new lineages of malware. (2) In real world settings, practitionners cannot be presented with a reliable dataset for training. Indeed, most malware are discovered, often manually, by antivirus vendors far later after they have been available to end-users [15]. Large-scale ML-based malware detection must therefore be used to automate the discovery of variants of malware which have been authenticated in a separate process.

Threat to Validity. To perform this study, we have considered a unique use-case scenario for using machine learning-based malware detection. This scenario consists in *Actively preventing malware from reaching markets* and is extremely relevant to most real-world constraints. Indeed, in practice, it is important to keep the window of opportunity very narrow. Thus, to limit the number of infected devices, Android malware must be detected as they arrive in the market. It is therefore important that state-of-the-art approaches be properly assessed, taking into account history constraints.

There is however a second use-case scenario which concerns online repositories for research and would consist on *cleaning such repositories regularly.* In this scenario, repositories maintainers attempt to filter malicious apps once a new kind of malware has been discovered. In such a context, practitionners can afford to wait for a long time before building relevant classifiers for identifying malware that have been since in the repository. Nevertheless, such repositories are generally of reasonable size and can be scanned manually and with the help of anti virus products.

There is a possibility that the results obtained in this paper would not be reproduced with either a different feature set and/or a different dataset. Nonetheless, we have no reason to believe that the way the dataset was collected induced any bias.

Future Work. (1) Building on the insights of our experiments, we plan to investigate how to maintain the performance of machine learning-based malware detectors until antivirus vendors can detect a new strain of malware. This research direction could help bring research work to be applied on real-world processes, in conjunction with antivirus products which are still widely used, although they do not scale to the current rates of malware production. (2) To account for the evolution of representations of malware lineages in training datasets over time, we plan to investigate a multi-classifier approach, each classifier being trained with more or less outdated data and weighted accordingly. A first challenge will be on how to infer or automate the choice of weights for different months in the timeline to build the most representative training dataset.

6   Conclusion

Given the steady increase in the adoption of smartphones worldwide, and the growth of application development for such devices, it is becoming important to protect users from the damages of malicious apps. Malware detection has thus been in recent years the subject of renewed interest, and researchers are investigating scalable techniques to spot and filter out apps with malicious traits among thousands of benign apps.

However, more than in other fields, research in computer security must yield techniques and approaches that are truly usable in real-world settings. To that end, assessment protocols of malware detection research approaches must reflect the practice and constraints observed by market maintainers and users. Through this empirical study we aim to prevent security research from producing approaches and techniques that are not in line with reality. Furthermore, given the performances reported in state-of-the-art literature of malware detection, while market maintainers still struggle to keep malware out of markets, it is important to clear the research field by questioning current assessment protocols.

In this paper, we have investigated the relevance of history in the selection of assessment datasets. We have performed large-scale experiments to highlight the different bias that can be exhibited by different scenarios of dataset selection. Our main conclusion is that the assessment protocol used for current approaches in the state-of-the-art literature is far from the reality of a malware detection practice for keeping application markets clean. We have further investigated naive approaches to training dataset construction and drawn insights for future work by the research community.

References

1. AppBrain: Number of available android applications, http://www.appbrain.com/stats/number-of-android-apps (accessed: September 09, 2013)
2. Allix, K., Bissyandé, T.F., Klein, J., Le Traon, Y.: A forensic analysis of android malware - how is malware written and how it could be detected? In: Proceedings of the IEEE Computer Software and Applications Conference, COMPSAC (2014)
3. Enck, W., Octeau, D., McDaniel, P., Chaudhuri, S.: A study of android application security. In: Proceedings of the 20th USENIX Conference on Security, SEC 2011, San Francisco, CA (2011)
4. Pieterse, H., Olivier, M.: Android botnets on the rise: Trends and characteristics. In: Proceedings of the Conference on Information Security for South Africa, ISSA (2012)
5. Idika, M.: A survey of malware detection techniques. Technical report, Purdue University (February 2007)
6. Arp, D., Spreitzenbarth, M., Hübner, M., Gascon, H., Rieck, K.: Drebin: Effective and explainable detection of android malware in your pocket. In: Proceedings of the Network and Distributed System Security Symposium, NDSS (2014)
7. Chau, D.H., Nachenberg, C., Wilhelm, J., Wright, A., Faloutsos, C.: Polonium: Tera-scale graph mining for malware detection. In: ACM SIGKDD Conference on Knowledge Discovery and Data Mining (2010)
8. Boshmaf, Y., Ripeanu, M., Beznosov, K., Zeeuwen, K., Cornell, D., Samosseiko, D.: Augur: Aiding malware detection using large-scale machine learning. In: Proceedings of the 21st Usenix Security Symposium (Poster session) (August 2012)
9. Su, X., Chuah, M., Tan, G.: Smartphone dual defense protection framework: Detecting malicious applications in android markets. In: Eighth IEEE International Conference on Mobile Ad-hoc and Sensor Networks, MSN (2012)
10. Henchiri, O., Japkowicz, N.: A feature selection and evaluation scheme for computer virus detection. In: Proceedings of the Sixth International Conference on Data Mining, ICDM 2006, Washington, DC, USA (2006)
11. Kolter, J.Z., Maloof, M.A.: Learning to detect and classify malicious executables in the wild. Journal of Machine Learning Research 7 (December 2006)
12. Zhang, B., Yin, J., Hao, J., Zhang, D., Wang, S.: Malicious codes detection based on ensemble learning. In: Xiao, B., Yang, L.T., Ma, J., Muller-Schloer, C., Hua, Y. (eds.) ATC 2007. LNCS, vol. 4610, pp. 468–477. Springer, Heidelberg (2007)
13. Sahs, J., Khan, L.: A machine learning approach to android malware detection. In: Proceedings of the IEEE European Intelligence and Security Informatics Conference, EISIC (2012)
14. Perdisci, R., Lanzi, A., Lee, W.: Mcboost: Boosting scalability in malware collection and analysis using statistical classification of executables. In: Proceedings of the Annual Computer Security Applications Conference, ACSAC (2008)
15. Apvrille, A., Strazzere, T.: Reducing the window of opportunity for android malware gotta catch 'em all. Journal of Computer Virology 8(1-2) (May 2012)
16. Alpaydin, E.: Introduction to Machine Learning, 2nd edn. The MIT Press (2010)

17. Cortes, C., Vapnik, V.: Support-vector networks. Machine Learning 20(3) (1995)
18. Breiman, L.: Random forests. Machine learning 45(1) (2001)
19. Cohen, W.W.: Fast effective rule induction. In: Proceedings of the International Machine Learning Conference. Morgan Kaufmann Publishers, Inc. (1995)
20. Quinlan, J.R.: C4. 5: programs for machine learning, vol. 1. Morgan Kaufmann (1993)
21. Pouik, G.: Similarities for fun & profit. Phrack 14(68) (April 2012), http://www.phrack.org/issues.html?id=15&issue=68
22. Cesare, S., Xiang, Y.: Classification of malware using structured control flow. In: Proceedings of the Eighth Australasian Symposium on Parallel and Distributed Computing, AusPDC 2010, vol. 107. Australian Computer Society, Inc., Darlinghurst (2010)
23. Allix, K., Bissyandé, T.F., Jerome, Q., Klein, J., State, R., Le Traon, Y.: Empirical assessment of machine learning-based malware detectors for android: Measuring the gap between in-the-lab and in-the-wild validation scenarios. Empirical Software Engineering (to be published, 2015)
24. Canfora, G., Mercaldo, F., Visaggio, C.A.: A classifier of malicious android applications. In: Proceedings on the 8th Conference on Availability, Reliability and Security (ARES) (2013)
25. Wu, D.J., Mao, C.H., Wei, T.E., Lee, H.M., Wu, K.P.: Droidmat: Android malware detection through manifest and api calls tracing. In: Proceedings of the 7th Asia Joint Conference on Information Security, AsiaJCIS (2012)
26. Amos, B., Turner, H., White, J.: Applying machine learning classifiers to dynamic android malware detection at scale. In: Proceedings of 9th International Wireless Communications and Mobile Computing Conference, IWCMC (2013)
27. Demme, J., Maycock, M., Schmitz, J., Tang, A., Waksman, A., Sethumadhavan, S., Stolfo, S.: On the feasibility of online malware detection with performance counters. In: Proceedings of the 40th Annual International Symposium on Computer Architecture, ISCA 2013, ACM, New York (2013)
28. Yerima, S., Sezer, S., McWilliams, G., Muttik, I.: A new android malware detection approach using bayesian classification. In: Proceedings of the 27th IEEE International Conference on Advanced Information Networking and Applications, AINA (2013)
29. Gascon, H., Yamaguchi, F., Arp, D., Rieck, K.: Structural detection of android malware using embedded call graphs. In: Proceedings of the 2013 ACM Workshop on Artificial Intelligence and Security, AISec (2013)
30. Bissyandé, T.F., Thung, F., Wang, S., Lo, D., Jiang, L., Réveillère, L.: Empirical Evaluation of Bug Linking. In: 17th European Conference on Software Maintenance and Reengineering (CSMR 2013), Genova, Italy (March 2013)
31. Jones, J.A., Harrold, M.J.: Empirical evaluation of the tarantula automatic fault-localization technique. In: Proceedings of the 20th IEEE/ACM International Conference on Automated Software Engineering, ASE. ACM (2005)
32. Hutchins, M., Foster, H., Goradia, T., Ostrand, T.: Experiments of the effectiveness of dataflow-and controlflow-based test adequacy criteria. In: Proceedings of the 16th International Conference on Software Engineering, ICSE (1994)
33. Rossow, C., Dietrich, C., Grier, C., Kreibich, C., Paxson, V., Pohlmann, N., Bos, H., van Steen, M.: Prudent practices for designing malware experiments: Status quo and outlook. In: 2012 IEEE Symposium on Security and Privacy (SP) (May 2012)
34. Böhme, R., Moore, T.: Challenges in empirical security research. Technical report, Singapoore Management University (2012)

35. Visaggio, C.A., Pagin, G.A., Canfora, G.: An empirical study of metric-based methods to detect obfuscated code. International Journal of Security & Its Applications 7(2) (2013)
36. Aafer, Y., Du, W., Yin, H.: DroidAPIMiner: Mining API-level features for robust malware detection in android. In: Zia, T., Zomaya, A., Varadharajan, V., Mao, M. (eds.) SecureComm 2013. LNICST, vol. 127, pp. 86–103. Springer, Heidelberg (2013)
37. Barrera, D., Kayacik, H., van Oorschot, P., Somayaji, A.: A methodology for empirical analysis of permission-based security models and its applications to android. In: Proceedings of ACM Conference on Computer and Communications Security, CCS (2010)
38. Chakradeo, S., Reaves, B., Traynor, P., Enck, W.: Mast: Triage for market-scale mobile malware analysis. In: Proceedings of ACM Conference on Security and Privacy in Wireless and Mobile Networks, WISEC (2013)
39. Peng, H., Gates, C.S., Sarma, B.P., Li, N., Qi, Y., Potharaju, R., Nita-Rotaru, C., Molloy, I.: Using probabilistic generative models for rangking risks of android apps. In: Proceedings of ACM Conference on Computer and Communications Security, CCS (2012)

Learning How to Prevent Return-Oriented Programming Efficiently

David Pfaff, Sebastian Hack, and Christian Hammer

CISPA, Saarland University, Saarbrücken, Germany
{pfaff,hack,hammer}@cs.uni-saarland.de

Abstract. Return-oriented programming (ROP) is the most dangerous and most widely used technique to exploit software vulnerabilities. However, the solutions proposed in research often lack *viability* for real-life deployment.

In this paper, we take a novel, statistical approach on detecting ROP programs. Our approach is based on the observation that ROP programs, when executed, produce different micro-architectural events than ordinary programs produced by compilers. Therefore, special registers of modern processors (*hardware performance counters*) that track these events can be leveraged to detect ROP attacks. We use machine learning techniques to generate a model of this different behavior, and develop a kernel module that detects *and prevents* ROP at runtime via the learned model. Our evaluation on real-world programs and attacks shows that the runtime overhead of this technique and the number false positives are very low, while preventing all known types of ROP attacks, including recently developed evasion techniques.

1 Introduction

The discovery of recent zero-day exploits against Microsoft Word [1], Adobe Flash Player [2] and Internet Explorer [3] demonstrate that return-oriented programming (ROP) is the most severe threat to software system security. Also, Microsoft's 2013 Software Vulnerability Exploitation trend report [4] found that 73% of all vulnerabilities are exploited via ROP. The core idea of ROP is to exploit the presence of so-called *gadgets*, small instruction sequences ending in a return instruction. By chaining gadgets together, an attacker is able to build complex exploits. The apparent popularity of ROP is explained by its power to bypass most contemporary exploit mitigation mechanisms, such as data execution prevention (DEP) [5] and address space layout randomization (ASLR) [6]. DEP and similar page-protection schemes prevent the execution of injected binary code, but ROP re-uses code already present in the executable memory segments, eliminating the need to inject code. ASLR randomizes the location of most libraries and executables, however, finding code segments left in a few statically known locations is often enough to leverage a ROP attack [7]. Since the inception of ROP by Shacham [8], research on ROP resembles an arms race:

F. Piessens et al. (Eds.): ESSoS 2015, LNCS 8978, pp. 68–85, 2015.
© Springer International Publishing Switzerland 2015

emerging defense techniques are continuously circumvented by increasingly subtle attacks [9,10,11].

In this paper, we take a novel, statistical approach on detecting ROP programs. Modern microprocessors spend most of their circuits on machinery that optimizes the execution of programs generated by compilers from "high-level" languages. Among this machinery are caches, translation look-aside buffers, branch predictors, and so on. To assist programmers in detecting performance problems, a modern CPU can record several hundred different kinds of micro-architectural events that occur during program execution (e.g. mispredicted branches, L1 cache misses, etc.). These events are counted by the CPU in special registers, the so-called *hardware performance counters* (HPCs).

In this paper, we claim *and experimentally verify* that the execution of a ROP program triggers such hardware events in a significantly different way than a conventional program that has been generated by a compiler. Essentially, micro-architectural events are a side channel by which a ROP program becomes distinguishable from a normal program at run time. There are several considerations that support this hypothesis: First, ROP programs use only indirect jumps (returns) to control the program flow. Common processor heuristics to detect the target of the return are useless in a ROP program because they do not follow the call/return policy. Second, ROP gadgets are small and scattered all over the code segment. Thus, there is no spatial locality in the executed code which should be observable in counters relevant to the memory subsystem.

We exploit the deviant micro-architectural behavior of ROP programs by training (using existing ROP exploits and benign programs) a support vector machine (SVM) based on profiles of hardware performance counters. Note, that despite our intuition we did *not* short-list any HPC types for training. We receive a classifier to distinguish ROP from benign programs and use it in a *monitor* kernel module that tracks the evolution of the performance counters and classifies them periodically. If the classifier detects a ROP program, defensive actions, like killing the process, can be taken.

We quantitatively evaluate the performance impact of HadROP on benign program runs using the SPEC2006 benchmark [12]: HadROP incurs a run time overhead of 5% on average and of 8% in the worst case. We also establish the effectiveness and practical applicability of HadROP in several case studies that show that HadROP detects and prevents the execution of:

- a ROP payload of an in-the-wild exploit on Adobe Flash Player [2].
- 25 new ROP payloads generated by the ROP-payload generator Q [13] that exploit manually injected vulnerabilities in GNU coreutils.
- Blind ROP [14] of an nginx web server. Using this recent dangerous technique, even amateurs can perform full-scale *remote-code execution* exploits.
- Multiple recent enhancements [9,10,11] that allow ROP to bypass previous hardware-assisted detection schemes. Our diversified monitoring scheme is *not* affected by those attacks in any practical scenario.

In summary, we make the following contributions:

- We present HadROP, a practical and easily deployable ROP defense mechanism, that does neither require instrumentation nor analysis of the code, be it source or binary.
- HadROP exploits the fact that ROP programs trigger micro-architectural events in modern micro-processors differently than conventional programs. Using state-of-the-art machine learning techniques, we train a classifier to detect these differences, which manifest themselves in the values of the HPC registers. HadROP's kernel-level run time monitor identifies ROP programs by periodically classifying the state of the HPCs.
- In several case studies, we evaluate HadROP on a set of existing and new ROP exploits for real-world applications. HadROP detected and prevented all exploits in our benchmark set without reporting a false positive. In a subsequent test of our kernel module on a production machine we encountered only three false positives in 24 hours. The performance overhead of HadROP is low, with an average of 5% on a set of computation-intensive benchmarks.

2   Return-Oriented Programming

Runtime attacks change the behavior of running processes. Typically, attackers exploit a buffer overflow vulnerability on the stack to overwrite the return address [15]. Pointing this address to any memory segment causes the segment to be interpreted as code, not data. By modifying the address to point towards a memory segment containing previously injected machine code, attackers are able to achieve arbitrary execution. Marking memory segments containing user-supplied data as non-executable prevents such code-injection attacks [5]. Many modern processors implement this protection mechanism in hardware in the form of an non-executable bit (NX-bit), which the OS can set for data memory pages.

ROP was developed as a response to circumvent protection mechanisms based on non-executable data pages [8]. For a ROP attack, the attacker overwrites the stack with a set of addresses pointing to already existing executable code. ROP carefully selects this set of addresses by finding suitable code locations to jump to. These code locations contain a few instructions and end on return, which is called a *gadget*. After executing the gadget, the return instruction will jump to the next address as specified on the stack. The addresses on the stack thereby effectively determines the program flow.

Figure 1 illustrates the mechanism of ROP. The payload on the stack (depicted on the the left) contains the addresses of multiple ROP gadgets. In addition to addresses, the payload may also contain parameters and other data that can be copied into registers. Upon execution of the `ret` instruction, the top-most address is interpreted as the return location. In this case, each return will jump towards the next gadget, thereby chaining the gadgets together. By combining appropriate gadgets, it is possible to incrementally build complex functions.

Initial ROP defense attempts primarily tried to prevent execution of malicious code, e.g. by monitoring programs at instruction level while checking for

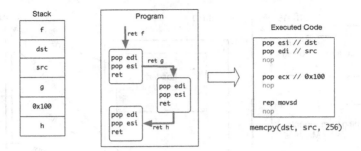

Fig. 1. Cross-section of a ROP exploit

an unusually high frequency of return-instructions [16]. However, often even small adaptations to the ROP paradigm, such as *jump-oriented programming* (JOP) [17], undermine the respective defense technique's core assumption. Therefore, other research focuses on hardening the program itself, e.g., by enforcing Control-Flow Integrity (CFI) [18] at runtime, which has been shown to be theoretically effective. In practice, the inclusion of code fragments for CFI severely degrades *performance* of the target program. Further, the required changes to the program either depend on access to *source code* (which is hard for proprietary software), or deteriorate *compatibility* both with other tools and runtime optimizations. For instance, just-in-time compilation (JIT) becomes impossible, because the generated JIT code does not include the required instrumentation, and thus violates the CFI property. As a result, the wide-spread deployment of those techniques becomes infeasible in practice [19].

3   Our Approach

Our approach builds on the fact that ROP programs differ significantly from conventional programs in terms of micro-architectural events in the CPU. Micro-architectural events are non-functional effects of the program execution on a modern CPU such as cache misses or mis-speculated branches. These events are essentially side channels that can be observed by the HPCs present in every modern CPU (cf. Section 3.1). However, it is hard to identify one particular kind of micro-architectural event that reliably indicates the execution of a ROP program in all its variants [9,10,11]. Therefore, in this work we determine a combination of different events that is characteristic for ROP execution. However, we do not start with a predefined conception of the kinds of events but assume that the exact combination of events *is dependent* on the CPU the program executes on, which is confirmed by our evaluation.

To this end, HadROP uses a statistical approach (cf. Fig. 2), that consists of an offline learning and an online monitoring component. The learning component records HPC profiles for a given set of ROP and conventional programs. Learning on the system the monitor is to be deployed on is imperative as other types of CPUs can lead to different HPC profiles. Then we leverage a SVM to obtain a

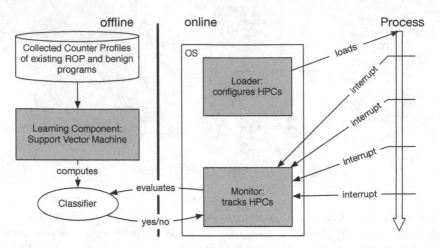

Fig. 2. Components of HadROP

classifier that discriminates ROP programs from benign programs. Once learned, this classifier remains constant on that particular CPU..

The monitoring component consists of a modified program loader and a kernel module. Whenever a new process is created, the modified program loader configures the CPU to track the set of HPCs the classifier needs to compute its result. Furthermore, it tells the CPU to raise an interrupt every N clock cycles. Upon every interrupt, the kernel module computes the difference of the current values of the HPCs to the values recorded at the previous interrupt and feeds those values to the classifier. If the classifier determines that an intrusion is about to happen, the process can be killed, or—depending on the application scenario—another defensive action can be taken such as notifying the user, security personnel, or a security information and event manager, potentially with additional information like a memory dump. Note that the HPCs are updated by the CPU itself, i.e., there is no performance overhead incurred by counting. The only overhead comes from handling the interrupt, reading the counters, and evaluating the classifier.

3.1   Hardware Performance Counters

The original purpose of HPCs is to non-intrusively profile a program by sampling the counters periodically. Usually, a CPU offers a small number of registers to count all sorts of micro-architectural events. On current CPUs 6–8 registers are freely configurable. To find a performance problem in the program, the programmer configures the HPCs to a certain set of events she deems relevant. Furthermore she sets a threshold per counter. If a counter reaches its threshold, the CPU resets the counter and raises an interrupt that is handled by the OS. Note that access to the HPCs is restricted to the operating system. A userspace program can never configure, read, or tamper with the HPCs. Therefore,

Table 1. Feature vector representation of a program run by collecting minimum and maximum of number of HPC events observed during a fixed period Δt

HPC event	\multicolumn{4}{c}{Program Run}	Feature Vector (max,min)
	Δt	Δt	Δt	Δt $\cdots$	
BR_MISP	5	1	4	21 $\cdots$	(21,1)
$\cdots$	$\cdots$	$\cdots$	$\cdots$	$\cdots$ $\cdots$	$\cdots$
ITLD_MISS	7	0	1	2 $\cdots$	(7,0)

in the case of profiling, the profiler needs to be a privileged application that communicates with the kernel to access the HPCs.

In general a CPU counts all configured events in the entire system because a CPU executes obliviously to the concept of a process. To break the counters down to individual processes they have to virtualized by the operating system: Each time the OS is entered the values of the HPCs are retrieved and stored in the process control block of the process that entered the OS. When the OS yields control to another process, it restores the counter values of this process from the respective process control block.

3.2 ROP Classification Using HPC Data

The purpose of the offline learning part as illustrated in Fig. 2 is to determine an unbiased model of which HPCs best indicate ROP attacks. Unbiased means that we make no pre-selection of a particular subset of HPC event types but treat them all equally. To determine this model, we use an SVM, a well-known machine learning technique. SVMs are beneficial here, because the resulting classifier is fast to evaluate, which is integral to our implementation. An SVM computes a hyperplane that separates a set of feature vectors into two clusters. In our setting, each feature vector captures the HPC profile of a particular program run. A feature vector consists of—for each event type—the minimum and maximum number of events per sampling period. Note that the sampling period has to be fixed before data collection starts and cannot be changed afterwards. Table 1 shows an example of two event traces of a program run and the corresponding feature vector. Remember from the last section that in practice this model can only refer to a small subset of the more than 200 event types available on a modern CPU. So in order to receive an unbiased model of the best combination, we sample event types in smaller batches and align the results after the fact. However, every measurement is noisy for various reasons: the complexity of the CPU makes it hard to reproduce the exact sequence of mirco-architectural events on every run. Furthermore, before the HPCs are saved to the process control block, the context switch might trigger additional events. Because every batch is collected individually, the feature vector of every batch is perturbed in a different way. Due to the large number of different HPC event types, an exhaustive process that trains and compares classifiers for each possible HPC combination is practically infeasible. As our experimental evaluation shows, this effect does not impede the ability of HadROP to successfully detect ROP programs.

Even so, our final classifier must adhere to the constraints of the target CPU, which means that we need to restrict the number of HPCs in our model to a small number that depends on the type of processor. This problem is well-known in machine learning as the *feature selection problem* [20]. Therefore, instead of learning a model based on all possible HPC event types, we leverage feature selection techniques to determine the, e.g., 16-event classifier that best matches the theoretical 200-event one in terms of predictive power.

After combining the batches into a feature vector for the full set of HPC event types, we have two kinds of feature vectors, those from ROP and from benign runs. To train a SVM classifier we feed the two sets of feature vectors into the learning algorithm, together with parameters like an error penalty C that allows more or less misclassification. Choosing the right parameters is not trivial, so we decided to learn these as well as part of our global optimization problem: *Which is the optimal model and subset of features that predicts ROP attacks best?* We solve this optimization problem using a search algorithm called *dynamic oscillating search* [21] to guide our feature selection algorithm. As the metric for this optimization problem we choose 10-fold cross validation, a standard model validation technique. In cross validation, the data set is partitioned into multiple subsets, of which a single subset is retained for testing and the remainder is used for training. This process is repeated until each set was retained at least once for testing. These evaluation results guide the dynamic oscillating search, which determines the next set of potentially optimal features and parameters. Iterating this process results in an error penalty and optimal[1] hyperplane based on a subset of HPC event types of appropriate size.

In practice, we use libSVM [22] to train the classifier, and the Feature Selection Toolbox [23] to provide the framework and wrapper for the dynamic oscillating search. Recall that we need to collect the HPC profiles of these program runs without recompiling or instrumenting these programs. To that end we replace the monitor in the kernel module presented in the next section with a module that writes the feature vectors to disk for offline classifier computation.

3.3 Data Collection

Producing a stable classifier requires a large set of benign and malicious (ROP-affected) behavior. To obtain a sufficiently large training set, we select several frequent targets of exploits in-the-wild *and* for which a large set of benign inputs are available. This gives us a good chance to collect reproducible exploits along with a number of samples of regular usage, as programs that are frequently exploited tend to be popular programs in general as well. We select Adobe Reader, mcrypt, PHP and the libtiff library to obtain a broad range of program usage scenarios. The programs and their associated vulnerabilities are outlined in Table 2. Exploits in-the-wild almost always focus on ROP and rarely use exotic variants such as jump-oriented programming or similar techniques. To increase the diversity of our training set we add own exploits based on these paradigms by adapting

---

[1] Due to the nature of such search algorithms one might potentially only determine a local optimum.

Table 2. Sample programs and their vulnerabilities used to train the ROP classifier

Program	Vulnerability	Description
Adobe Reader 9.3	CVE-2010-0188, CVE-2006-3459	Exploits integer overflow in libtiff library when rendering TIFF images. Used as case study for ROPdefender [25].
tiffinfo 3.9.2	CVE-2010-2067	Stack-based buffer overflow due to insufficient bounds checking during processing of EXIF tags [36].
PHP 5.3.3	CVE-2011-1938	Stack-based buffer overflow in the socket_connect function. Used as case study for Microgadgets [37].
mcrypt 2.6.8	CVE-2012-4409	Stack-based buffer overflow in command line utility for encrypted file headers with overly long salt data.

Table 3. Payloads used for different exploit variants

Variant	Description
ROP-only payload	Fully consists of gagets and does not use libc functions or injected code segments.
Jump-oriented Programming	ROP-like code-reuse attack whose gadgets end on indirect jumps instead of return [17].
Return to mprotect	ROP chain loads ordinary shellcode payload into memory, then calls mprotect to make it executable.
Multi-stage payload (stack pivoting)	Builds a fixed stack in data section with ROP, then execute normal ROP payload.

the in-the-wild exploits. Details of the types of payload variants are outlined in Table 3. On average, these payloads take 2390 instructions to execute the complete ROP shellcode. The samples range from 200 (a libtiff exploit) at the lowest to over 6000 (an Adobe Reader exploit) at the highest. In addition to malicious input, we also collect benign input from a number of sources: We collect benign PDF files from the arXiv archive [2] and the Intel Software Developer Manuals [3], benign PHP source files from GitHub trending PHP repositories [4], benign TIFF images from the sample sources of the libtiff library [5] and benign usage samples of mcrypt by selectively encrypting the other samples.

3.4 Kernel Monitor

The active, online component of HadROP is implemented as a patch to the Linux kernel. It consists of three main components: a modified program loader, an extended interrupt handler, and changes to the HPC API.

---

[2] http://arxiv.org/
[3] http://www.intel.com/content/www/us/en/processors/architectures-software-developer-manuals.html
[4] http://github.com/
[5] http://www.remotesensing.org/libtiff/

The program loader is modified in the routines that start a program. It configures the HPCs needed for the classifier in the monitor. Currently, the configuration is the same for all processes, but virtualized for every thread. This means that there is no interaction between the program loader and the monitor beyond the initial configuration of the HPCs. After initializing the HPCs, the program loader proceeds as usual. Note that it is possible to disable this step by preloading a custom program loader routine using LD_PRELOAD. This is not a security issue, but a feature of our approach. Since preloading is disabled for security-critical programs (such as setuid applications), we can give unprivileged applications the ability to disable this protection mechanism. As a result, HPCs can be configured by programs and their original purpose, performance evaluations during program development, is preserved.

The interrupt handler in perf_events is patched to recognize the configuration specific to the classifier. Configurations not specific to HadROP are handled as usual to preserve the original HPC functionality. Interrupts produced by HPCs are recognized as non-maskable interrupts, which must be handled by the kernel and cannot be ignored (masked). The monitor containing the classifier is called from within the interrupt handler. It reads the contents of all relevant HPCs and applies the SVM classifier on the values. Depending on the outcome of the classifier, it either terminates the interrupt handler to pass control back to user mode, or executes a defensive action, e.g. kills the process.

Finally, we change the API that exposes the HPC to user mode. In particular, we need to restrict processes from modifying their own HPC configuration during observation of HadROP. This restriction is necessary, since otherwise, ROP-affected processes could simply mask their execution by changing the HPC configuration.

4 Evaluation

Our main evaluation environment consists of an Intel Core i7-4800MQ 3.7GHz system running Linux Mint 15 OS with Linux kernel version 3.11 The evaluation is partitioned into multiple experiments that validate different aspects of HadROP.

4.1 Learning Environment and Classifier

The measurement period has to be determined experimentally *before* training the classifier: since the period affects the magnitude of the measurements, we cannot treat it as another parameter to be optimized during the feature selection process. Therefore we repeat the data-collection and the optimization process multiple times for a selection of different values. We compare the accuracy and slowdown of those experimental runs in Figure 3.

4.2 Evaluation of HPCs Chosen by the SVM

An evaluation of the classifier entails a thorough analysis of the HPCs used within the classifier. In particular, we want to answer the following questions:

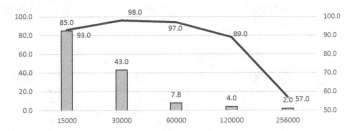

Fig. 3. Trade-off between slowdown (bars) and overall detection accuracy (line)

Table 4. Evaluation of our trained classifier

(a) Accuracy of classifiers trained on CPU X for detection on CPU Y.

trained \ used	CPU 1	CPU 2
CPU 1	97.3	83.9
CPU 2	79.5	98.1
CPU 1&2	78.1	79.6

(b) SPEC2006 Integer Benchmark slowdown induced by our kernel patch.

Benchmark	Slowdown	Benchmark	Slowdown
perlbench	4.98%	sjeng	6.60%
bzip	4.60%	libquantum	7.67%
gcc	7.32%	h264ref	7.32%
mcf	4.95%	omnetpp	4.84%
gobmk	8.06%	astar	4.62%
hmmer	4.88%	xalancbmk	4.53%

Do multiple HPCs event type improve ROP detection? Is the learned information generic or specific to each processor?

To answer the first questions we analyze the distributions of individual HPC events for benign and malicious executions. Figure 4 shows *box-and-whisker* plots of both (normalized) distributions. We observe that the ROP-affected runs display little variance. In contrast, benign executions cover a larger portion of the space. Most importantly, the outliers of benign execution stretch across the region occupied by malicious execution. Hence we cannot reasonably expect to infer ROP-execution from only a single measurements without an unreasonably high false-positive rate. Note that benign execution only shows these outliers on a small subset of HPC events at the same time. Finding an appropriate subset that minimizes the correlation of benign outliers maximizes the indicativeness for ROP-affected behavior. Our automated approach of fine-tuning and optimizing an SVM classifier using feature-selection shows much better performance than any singular measurement.

To answer the second question we take an experimental approach and compare the classifiers of two different processors, CPU1 (Core i7-4800MQ 3.7GHz) and CPU2 (Core i7 860 3.4GHz). Table 4a shows that using the classifier trained on CPU1 for runs on CPU2 decreases the detection rate significantly. Vice versa, the converse yields similar results. Even training the classifier on both CPU1 and CPU2 using a combined data set does not improve the detection rate for either CPU.

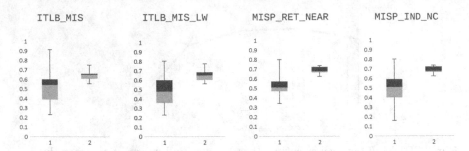

Fig. 4. Distribution of chosen HPCs for benign (1) and ROP-affected (2) runs

4.3   Performance Impact

To assess the overall performance impact of HadROP's monitor to compute-intensive applications, we ran the integer benchmarks from the standard performance benchmark suite SPEC2006 [12]. For each benchmark, we measured the wall-clock-time and compared it to runs during which the monitor was not active. The results shown in Table 4b illustrate that we have an average overhead incurred by HadROP of about 5% and a maximum of 8%. This overhead consists of the context-switch costs, the handling of the interrupt, and the evaluation of the classifier. The largest share of this overhead can be attributed to the context-switch costs. None of these (non-trivial) benchmarks raises a false positive.

4.4   Obtrusiveness

An important part in the viability of monitoring schemes is how much they negatively impact everyday usage scenarios. While false positive rates of individual classifiers are useful to give a rough idea on the relative performance, they can and should not be treated as absolutes. Hence we conducted an experiment, wherein patched systems are challenged with typical usage scenarios in a time frame of 24 hours. Typical usage includes browsing the internet on different browsers, coding and debugging and watching videos and listening to audio files. In the course of 24 hours we were notified of only three false positives. In all cases, these false positives happened close to system (re-) starts: twice within a debugger and once during an automated software update. Dumps of the HPC data showed that during start-up periods more misses and mispredictions were recorded, although the impact is usually not severe enough to be reported as a false positive.

4.5   Case Study: Adobe Flash Player

The purpose of this case study is to evaluate the effectiveness of HadROP on attacks that were not part of the training dataset. We ran (an adaption of) a recently discovered Adobe Flash Player exploit [2]. In the wild, this attack was targeting Windows and OS X systems, but the vulnerability and general procedure is applicable to Linux as well. The attack includes several advanced ROP techniques, such as *stack pivoting*, which compiles a ROP-stack in an attacker-controlled buffer on the heap. We replaced the final code as to open a root shell and verified that the attacks succeeds on an unpatched system.

Afterwards, we patched the system with our kernel module. Note that the classifier had not been trained on Adobe-Flash specific HPC data (cf. Section 3.3). As expected, playing benign swf files was *not* classified as a ROP attack. However, opening the malicious swf file, the player was killed by our kernel module before a root shell could be opened. We conclude that our approach is therefore effective even for programs whose behavior was not analyzed during the offline training stage.

4.6   Case Study: Automatically Derived ROP Payloads

In this case study, we demonstrate that HadROP is able to detect very *diverse* ROP programs that we generated with state-of-the-art tools. This is conclusive evidence that HadROP's classifier is not overfitted to the training set and generalizes well. ROP payload generators such as Q [13] automatically derive payloads using semantic program verification techniques to identify gadget functionality. Similar to ROPecker [29], we automatically generate payloads for binaries in usr/bin/ and bin/. However, while in ROPecker the authors simulate the exploitation, we manually modified a subset of the binaries to mimic a realistic vulnerability as faithful as possible. Altogether, we modify 25 binaries such that they call a vulnerable function leading to a buffer overflow or enable a function pointer overwrite. We then use Q to automatically derive ROP payloads for those binaries. Due the restricted number of gadgets available to Q in these binaries, it is forced to frequently employ different gadget chains and strategies. This yields a diverse set of gadget behavior beyond the ones frequently employed for the exploitation of programs with a large code base. Because those small programs provide only little code that is exploitable by ROP, they pose an additional challenge to HadROP: TLB-related counters are not triggered as often as for programs with large code bases. Because HadROP relies on a *combination* of counters, it can compensate for this effect. For all executions of ROP payloads generated by Q, HadROP intervened. Hence we have evidence to claim that HadROP can even detect previously unseen ROP-attacks even if the exploits have a different character than those used to train the classifier.

4.7   Case Study: Nginx and Blind ROP

Our final case study is concerned with one of the most dangerous scenarios, in which a remote service can be exploited without knowledge of the binary. While automated ROP payload derivation tools such as Q already make it easy for a semi-experienced hacker to build a functional shellcode, Blind ROP (BROP) [14] only requires minimal knowledge and can easily be distributed to amateurs.

A typical BROP attack proceeds in multiple stages. First it searches for the base address of the executable. During all guessing stages, the attacker monitors whether the exploited application crashed due to an invalid memory address. This process will eventually lead to valid addresses. In the following stages, the exploit searches for specific types of gadgets as ROP building blocks. Afterwards, it identifies two useful functions that dump the entire binary in order to derive the final ROP chains for the binary on-the-fly. Note that this exploit consists of two separate ROP chains: one to search for gadgets and to dump the binary and one to execute the real exploit built from the dumped binary.

For our evaluation, we apply this attack to the nginx web server vulnerability CVE-2013-2028. We host the appropriate nginx web service on our machine and configure it to constantly restart after a crash, to fulfill BROPs requirements. Afterwards we start the attack remotely. In our experiments, HadROP detected all malicious queries and the resulting ROP-execution reliably and terminated the web server before the final ROP chain could be fully executed. Note that we performed further case studies like the attacks proposed in [9,10,11] (see discussion in Section 6), for which we do not have space to discuss them in detail here.

5   Related Work

Runtime Monitoring. Runtime solutions usually monitor execution at the instruction level. A sophisticated method to achieve transparent monitoring is the use of dynamic binary instrumentation frameworks such as Pin [24]. Dyn-IMA [16] are exemplary of early runtime monitor attempts to detect the execution of ROP exploits by checking for an abnormally high frequency of return instructions. ROPDefender [25] instruments call and ret instructions to update a separately maintained shadow stack; ROP is then detected by comparing the shadow stack to the actual system stack on every return. All those early tools do no detect jump-oriented programming and similar ROP variants that do not use `ret` gadgets. ROPStop [26] is a sophisticated instrumentation based on a similar observation as ours: regular, compiler-produced programs adhere to a set of normal execution behavior, while ROP executes chains of short gadgets without regarding their location. However, they differ from us in monitoring these properties by analyzing the program binary, including the extraction of a control flow graph, at system-call entries. Like similar methods, this approach is dependent on accurate and fast disassembly. ROPGuard [27] provides a set of system-call wrappers that verify a set of properties on `call`, such as the instruction preceding the return address being a call instruction. This indicates that the return is targeting a genuine return address.

Hardware-assisted Program Monitoring. A major drawback of most runtime monitoring solutions is their dependency on program instrumentation and high runtime overhead. Recently, multiple solutions surfaced which leverage existing hardware components to enforce security properties. kBouncer [28] utilizes the

Last Branch Record (LBR) facilities of modern processors, which record the branch targets of the last 4-16 branch instructions. They verify the validity of program execution by running heuristics on LBR data upon system API entry. ROPecker [29] improves on the idea of kBouncer by not only inspecting the previous indirect branch targets, but also simulating future ones using elaborate prediction mechanisms based on LBR data. Yuan et al. [30] pre-analyze the binary and store instructions preceding return instructions. It was one of the first works to identify HPC events as indicators for code injections attacks, but not ROP. Wicherski [31] successfully links a single branch prediction event to ROP execution. In contrast to HadROP, all of these systems have been shown to be susceptible to attacks [9,10,11]. We consider their focus on a single type of feature as the root cause to their vulnerabilities.

Another security application of HPC was demonstrated by Malone et al. [32], who pre-record HPC data profiles for individual binaries that can be compared to the execution traces at runtime. This approach successfully finds minor deviations in runtime behavior, indicating an integrity breach. Demme et al. [33] use the same technique to extract malware signatures. Both techniques are focused on intrusion detection instead of attack prevention and therefore orthogonal to our approach. Furthermore, we do restrict ourselves to the detection of ROP, but do not need to analyze the binary. As a result, we manage to capture an intrinsic property of ROP in its effects on HPCs across different programs.

Concurrent work [34] presents a system to detect drive-by attacks in Internet Explorer (and some plugins). While it is close to our research in terms of technology (leveraging machine learning and HPCs) they concentrate their detection on "phase1" after a ROP attack has already disabled DEP. This reflects in their results: They can detect malware in "phase1" with very high success rate, but in the ROP phase their detection rate is low. In contrast, our work targets ROP purely, in all of its variants (even if they do not disable DEP) and for all programs on a system, which led to the development of a kernel module.

6 Discussion

As noted in Section 3.1, modern CPUs only make a limited amount of HPCs available, which are further restricted by non-documented conflicts at measuring related HPC events. Given that previous generations of processors provided more registers for concurrent use, this is a negative development. Furthermore, the accuracy of measurements on the HPCs are subject to noise, i.e., the counters are not required to perfectly correspond to the executed program. Improvements in these areas might improve detection accuracy due to measuring less noisy and more diverse event data, which reduces the number of false positives.

The integration of machine learning techniques also incurs machine-learning specific limitations. Our classifier was trained on a set of data, which might not be representative of all possible exploits. In the process of the classifier derivation, we analyzed roughly 40 exploits based on ROP and its variants. All exploits were based on real-world exploits and adapted to a Linux operating system and variants of ROP exploits that are not encountered in-the-wild (JOP). This learned classifier thus does not fully guarantee that we will immediately detect new, as of yet unseen, ROP variants even though we have successfully prevented a large set of untrained attacks. In the event that a ROP variant is not detected, however, we can include it in our analysis to produce a classifier capable of detecting that class of exploits in the future.

The behavior-based detection of HadROP supports an easy adaption to all currently employed ROP variants, yet it potentially makes it susceptible to attacks that try emulate benign behavior—so-called *mimicry attacks*. We evaluate the feasibility of two possible variants of such attacks in the following. *Call-preceded gadgets* [9] attack the assumption that gadgets do not form call-return pairs. As ROPecker and kBouncer do not determine whether the return-location corresponds to the correct call, their defense mechanisms only limits the number of useful gadgets. In contrast, TOOL is not affected by call-preceded gadgets, as the CPUs return-prediction requires actual call-return pairs. *History flushing* targets defense mechanisms that only maintain a limited history, such as storing the last 20 branch targets. *Evasion attacks* operate similarly, but instead they evade detection immediately by only using gadgets that directly contradict the defense's core assumptions. In our context, an attacker attempting to evade detection must actively manipulate *all* relevant HPC values for the execution to seem benign. This process is far from trivial for an LBR record, for multiple HPCs this becomes onerous. For the target binaries for which history-flushing and evasion attacks could be constructed, our tests show that these attacks eventually become powerful enough to evade detection, but this comes at a price. Using meaningless but inconspicuous computations to cloak the ROP exploit leaves an attack with less than 10% effective computation progress as noted in [9]. When testing this attack pattern to enhance the original flash exploit from subsection 4.5 we even observed a 20-times slowdown. Thus, the ROP attack does not finish within multiple minutes, during which the process will be unresponsive. Hence we believe that such attacks cannot be feasibly used in practice to attack TOOL. Note that all of these studies were performed on a system with ASLR disabled. Other recent orthogonal techniques (e.g. [35]) render the construction of these kinds of attacks even more infeasible.

7   Conclusion

ROP exploits are the most severe threat to software security. Traditional techniques that have shown to be effective at preventing ROP require the analysis or instrumentation of source or binary code, which severely hampers their deployment. Our approach is novel in using a statistical approach that observes the micro-architectural events of a CPU, which are abnormal for ROP programs.

We detect ROP programs with very high confidence: In our experiments we detected and prevented *all* of our real-world benchmark ROP attacks, be it known exploits or newly-crafted attacks. Our technique produced *no* false positives on all benchmark programs and only three false positives in a 24 hour usage test but—more importantly—*no false negatives.* The run time monitor incurs 5% slow down on average and 8% at maximum. In contrast to related techniques, we neither require analysis nor modification of the application source or binary code. Our implementation as a kernel patch makes our system easy to deploy to production systems.

Acknowledgments. This work was supported by the German Ministry for Education and Research (BMBF) through funding for the Center for IT-Security, Privacy and Accountability (CISPA).

References

1. Emerging threat: Microsoft word zero day (cve-2014-1761) remote code execution vulnerability,
   http://www.symantec.com/connect/blogs/emerging-threat-microsoft
   -word-zero-day-cve-2014-1761-remote-code-execution-vulnerability
2. Security updates for adobe flash player,
   http://helpx.adobe.com/security/products/flash-player/apsb14-07.html
3. Vrt: Anatomy of an exploit: Cve 2014-1776,
   http://vrt-blog.snort.org/2014/05/anatomy-of-exploit-cve-
   2014-1776.html
4. Microsoft: Software vulnerability exploitation trends (2013)
5. van de Ven, A.: New security enhancements in red hat enterprise linux v.3, update 3. Technical report, Red Hat, Raleigh, North Carolina, USA (2004)
6. The PaX Team, https://pax.grsecurity.net/
7. Snow, K.Z., Monrose, F., Davi, L., Dmitrienko, A., Liebchen, C., Sadeghi, A.R.: Just-in-time code reuse: On the effectiveness of fine-grained address space layout randomization. In: 34th Symposium on Security and Privacy (SP), 574–588. IEEE (2013)
8. Shacham, H.: The geometry of innocent flesh on the bone: Return-into-libc without function calls (on the x86). In: 14th Conference on Computer and Communications Security, pp. 552–561. ACM (2007)
9. Carlini, N., Wagner, D.: Rop is still dangerous: Breaking modern defenses. In: 23rd USENIX Security Symposium, San Diego, CA, pp. 385–399 (2014)
10. Davi, L., Sadeghi, A.R., Lehmann, D., Monrose, F.: Stitching the gadgets: On the ineffectiveness of coarse-grained control-flow integrity protection. In: 23rd USENIX Security Symposium, San Diego, CA, pp. 401–416 (2014)
11. Göktaş, E., Athanasopoulos, E., Polychronakis, M., Bos, H., Portokalidis, G.: Size does matter: Why using gadget-chain length to prevent code-reuse attacks is hard. In: 23rd USENIX Security Symposium, San Diego, CA, pp. 417–432 (2014)
12. Spec standard performance evaluation corporation, http://www.spec.org
13. Schwartz, E.J., Avgerinos, T., Brumley, D.: Q: Exploit hardening made easy. In: 20th USENIX Security Symposium, San Francisco, CA (2011)
14. Bittau, A., Belay, A., Mashtizadeh, A., Mazieres, D., Boneh, D.: Hacking blind. In: 35th Symposium on Security and Privacy (S&P), vol. 14. IEEE (2014)
15. Aleph One: Smashing the stack for fun and profit. Phrack Magazine 7 (1996)
16. Davi, L., Sadeghi, A.R., Winandy, M.: Dynamic integrity measurement and attestation: towards defense against return-oriented programming attacks. In: Workshop on Scalable Trusted Computing, pp. 49–54. ACM (2009)

17. Bletsch, T., Jiang, X., Freeh, V.W., Liang, Z.: Jump-oriented programming: a new class of code-reuse attack. In: 6th Symposium on Information, Computer and Communications Security, pp. 30–40. ACM (2011)
18. Abadi, M., Budiu, M., Erlingsson, U., Ligatti, J.: Control-flow integrity. In: 12th Conference on Computer and Communications Security, pp. 340–353. ACM (2005)
19. Szekeres, L., Payer, M., Wei, T., Song, D.: Sok: Eternal war in memory. In: 34th Symposium on Security and Privacy (S&P), pp. 48–62. IEEE Computer Society, Washington, DC (2013)
20. Guyon, I., Elisseeff, A.: An introduction to variable and feature selection. The Journal of Machine Learning Research 3, 1157–1182 (2003)
21. Somol, P., Novovicova, J., Grim, J., Pudil, P.: Dynamic oscillating search algorithm for feature selection. In: 19th International Conference on Pattern Recognition, pp. 1–4 (2008)
22. Chang, C.C., Lin, C.J.: LIBSVM: A library for support vector machines. ACM Transactions on Intelligent Systems and Technology 2, 27:1–27:27 (2011)
23. Feature selection toolbox, http://fst.utia.cz/
24. Luk, C.K., Cohn, R., Muth, R., Patil, H., Klauser, A., Lowney, G., Wallace, S., Reddi, V.J., Hazelwood, K.: Pin: building customized program analysis tools with dynamic instrumentation. ACM Sigplan Notices 40, 190–200 (2005)
25. Davi, L., Sadeghi, A.R., Winandy, M.: Ropdefender: A detection tool to defend against return-oriented programming attacks. In: 6th Symposium on Information, Computer and Communications Security, pp. 40–51. ACM (2011)
26. Jacobson, E.R., Bernat, A.R., Williams, W.R., Miller, B.P.: Detecting code reuse attacks with a model of conformant program execution. In: Jürjens, J., Piessens, F., Bielova, N. (eds.) ESSoS 2014. LNCS, vol. 8364, pp. 1–18. Springer, Heidelberg (2014)
27. Fratric, I.: Runtime prevention of return-oriented programming attacks (2012), http://ropguard.googlecode.com/
28. Pappas, V., Polychronakis, M., Keromytis, A.D.: Transparent rop exploit mitigation using indirect branch tracing. In: 22nd USENIX Security Symposium (2013)
29. Cheng, Y., Zhou, Z., Yu, M., Ding, X., Deng, R.H.: Ropecker: A generic and practical approach for defending against rop attacks. In: The 21st Annual Network and Distributed System Security Symposium (2014)
30. Yuan, L., Xing, W., Chen, H., Zang, B.: Security breaches as pmu deviation: detecting and identifying security attacks using performance counters. In: Second Asia-Pacific Workshop on Systems, p. 6. ACM (2011)
31. Wicherski, G.: Taming rop on sandy bridge. Syscan (2013)
32. Malone, C., Zahran, M., Karri, R.: Are hardware performance counters a cost effective way for integrity checking of programs. In: Sixth Workshop on Scalable Trusted Computing, pp. 71–76. ACM (2011)
33. Demme, J., Maycock, M., Schmitz, J., Tang, A., Waksman, A., Sethumadhavan, S., Stolfo, S.: On the feasibility of online malware detection with performance counters. In: 40th Annual International Symposium on Computer Architecture, pp. 559–570. ACM (2013)
34. Tang, A., Sethumadhavan, S., Stolfo, S.J.: Unsupervised anomaly-based malware detection using hardware features. In: Stavrou, A., Bos, H., Portokalidis, G. (eds.) RAID 2014. LNCS, vol. 8688, pp. 109–129. Springer, Heidelberg (2014)
35. Backes, M., Nürnberger, S.: Oxymoron: Making fine-grained memory randomization practical by allowing code sharing. In: 23rd USENIX Security Symposium, San Diego, CA, pp. 433–447 (2014)

36. Rosenberg, D.: Breaking libtiff,
http://vulnfactory.org/blog/2010/06/29/breaking-libtiff/
37. Homescu, A., Stewart, M., Larsen, P., Brunthaler, S., Franz, M.: Microgadgets:
Size does matter in turing-complete return-oriented programming. In: WOOT,
pp. 64–76 (2012)

A   HPCs Chosen by the SVM

Table 5 shows the events our machine learning technique determined to be the
most indicative of ROP computations. Intuitively, we can categorize these event
types in roughly three classes: First, *instruction translation look-aside buffer*
(ITLB) misses: ROP gadgets are short and jump around "wildly" in memory,
which breaks the common spatial locality properties of conventional code. Furthermore, they use code snippets in libraries that the original program might
not have executed at all. Hence, an increase in ITLB misses is not unexpected.
Full ITLB misses lead to a computationally very expensive page table lookup, in
a process commonly referred to as a *page walk*. Second, events related to branch
prediction. Indirect jumps executed by ROP programs are inherently unpre-
dictable because they do not follow the common call/return pattern and ROP
programs rarely execute loops that could increase the prediction performance.
Third, stalls in the instruction pipeline, which most likely results from frequent
prediction or cache misses. Note that while these types of events might seem
natural candidates, other types of processors might induce a different optimal
subset of HPC event types for ROP detection. An evaluation on the structurally
very similar i7 860 already shows differences in the selection of the events. While
the branch prediction related events remained the same, ITLB-related events
differ strongly in one event. Note that the actual functionality of HPC events
might differ across processors, even for events of the same mnemonic.

Table 5. HPCs in the classifier for Intel i7 4800MQ (above) and i7 860 (below)

ITLB_MISSES.MISS_CAUSES_WALK	Miss in instruction translation look-aside buffer (ITLB)
ITLB_MISSES.LARGE_WALK_COMPLETED	ITLB miss causing complete page walk
BR_MISP_EXEC.INDIRECT_NON_CALL.TAKEN	Mispredicted indirect non call branch executed
BR_MISP_EXEC.RETURN_NEAR.TAKEN	Mispredicted return branch executed
L1I.CYCLES_STALLED	Stall in L1I Cache due to ITLB fault or miss
ITLB_MISSES.WALK_COMPLETED	ITLB miss causing complete page walk
BR_MISP_EXEC.INDIRECT_NON_CALL.TAKEN	Mispredicted indirect non call branch executed
BR_MISP_EXEC.RETURN_NEAR.TAKEN	Mispredicted return branch executed

Re-thinking Kernelized MLS Database Architectures in the Context of Cloud-Scale Data Stores

Thuy D. Nguyen, Mark Gondree, Jean Khosalim, and Cynthia Irvine

Department of Computer Science, Naval Postgraduate School
Monterey, California 93943
{tdnguyen,mgondree,jkhosali,irvine}@nps.edu

Abstract. We re-evaluate the kernelized, multilevel secure (MLS) relational database design in the context of cloud-scale distributed data stores. The transactional properties and global integrity properties for schema-less, cloud-scale data stores are significantly relaxed in comparison to relational databases. This is a new and interesting setting for mandatory access control policies, and has been unexplored in prior research. We describe the design and implementation of a prototype MLS column-store following the kernelized design pattern. Our prototype is the first cloud-scale data store using an architectural approach for high-assurance; it enforces a lattice-based mandatory information flow policy, without any additional trusted components. We highlight several promising avenues for practical systems research in secure, distributed architectures implementing mandatory policies using Java-based untrusted subjects.

1 Introduction

Resource sharing exists at several layers in the cloud. For example, platform-as-a-service usually employs virtualization with shared hardware; software-as-a-service may provide multi-tenant database services (*e.g.*, [37]). Given this shared resource environment, information leakage is a major concern in a multi-customer cloud [30,26]. Further, a variety of sensitive data is being managed by community and private clouds in governments and industries across the globe, *e.g.*, healthcare organizations in the U.S. and EU [12]. The U.S. government is using a community cloud infrastructure for processing and sharing intelligence data [25] and is planning different tactical cloud environments [6,24] for ground and afloat operations. The output of tactical sensors to these clouds has been dubbed a "Data Flood" problem [29]. These Big Data challenges go beyond the need for new analytics: a leak resulting from this flood may pose grave danger to both human intelligence sources and national security. The need to manage sensitive and classified data in shared cloud infrastructures motivates enforcing strict, mandatory policies for information flow control, through the use of systems following the same rigor applied to a security kernel [32].

F. Piessens et al. (Eds.): ESSoS 2015, LNCS 8978, pp. 86–101, 2015.

It is in this context that we re-explore the design of high-assurance multi-level secure database systems, adapted for cloud-scale data stores. Prior work has considered relational database management systems (RDBMS) preserving mandatory information flow policies (*i.e.*, the Bell-LaPadula model). At the time, relational databases appeared to be a multi-purpose "one size fits all" solution; this perspective, however, has substantially waned [35]. An emerging trend is to select the data model and query model one's application requires, then to select a storage back-end appropriate for the situation. Experience has begun to show that often the resultant model is not relational, nor does the application require ACID properties (atomicity, consistency, isolation, durability). As a result, various high-availability, massively-scalable, non-relational ("NoSQL") databases have found adoption in a cloud context. These new databases tend to not guarantee ACID properties, instead relaxing consistency in favor of availability and network partition tolerance. The success of non-relational databases to support a variety of cloud services has demonstrated that many natural and important applications—*e.g.*, content distribution, content management systems, massively parallel data mining—are not efficiently maintained as relational models.

Our work makes the following contributions:

- We formulate the problem of multilevel security for cloud data stores—prior work considered only MLS relational models and MLS transactional systems.
- We propose the design of a scalable data store following BigTable's design, capable of enforcing an MLS policy; the design uses a variant of the kernelized architecture approach, requiring no trusted components external to the OS.
- We implement a prototype of our design using Apache HBase and HDFS, requiring only small modification to run as MLS-aware untrusted subjects.
- We experimentally evaluate our prototype, verifying compatibility using several popular cloud applications (e.g., Titan, Apache Storm) and assessing performance using known cloud benchmarking tools.

We identify limitations in adapting a large class of cloud applications—*i.e.*, those making extensive use of in-memory data structures, employing languages like Java—for kernelized systems. Our performance experiments call into question the viability of the kernelized design in the context of cloud-scale systems; we discuss these findings and suggest possible directions for future work.

Our prototype follows the Hinke-Schaefer variant of the kernelized database architecture pattern. This design pattern allows the entire application to be executed without privileges in an MLS environment, while supporting all allowable access patterns, *e.g.*, read-down. This is motivated by trusted computing base (TCB) minimization requirements [1,23]. Untrusted applications built around this pattern are called *MLS-aware* [21]. We select this MLS database architecture for exploration as it is credited as best facilitating the "retrofit" of existing code to run on high-assurance systems [15]. It is known to be inefficient when tuple-level labels are required, as data must be divided among many different operating system objects [17,15]; thus, our prototype only supports labels at the coarsest (table-level) granularity. From a security perspective, the only major criticism of Hinke-Schaefer relates to support for transactions, which our design avoids by adopting a weakened consistency model.

2   Related Work

The problem of information flow control has received growing attention in infrastructure-as-a-service (IaaS) cloud service models. In particular, some projects have explored the threat of placing co-resident VMs in shared clouds [30,5] for side-channel attacks [39,40]. The Xenon VMM [26] is a hardened version of Xen satisfying a separation policy appropriate for controlling these flows. Relatedly, Wu *et al.* [38] design a proof-of-concept IaaS system based on Eucalyptus, implementing Chinese Wall rather than a strict separation policy. Watson [36] proposes a more general set of rules for information flow control between sets of nodes performing a joint computation in a federated setting. Information flow control in storage-as-a-service models, however, has not been well-explored, nor have the lessons of MLS RDBMS research been re-evaluated in this new domain.

Some non-relational data stores support native mechanisms for access control. Apache Accumulo [2] is a column-store that extends the BigTable design to support cell-level access control. Each cell is assigned a security label encoding non-discretionary, attribute-based access control rules; these are not equivalent to MLS labels, *i.e.* they are not used to enforce an information flow control policy. In particular, users with permissions to write a cell can modify its label, or write this data to a new cell with less restrictive visibility (in MLS terms, either violating tranquility or performing a downgrade). Relatedly, Apache HBase implements access control lists (ACLs) at the table- and column-granularity. As of v0.98, HBase features both cell-level visibilities, like Accumulo, and ACLs on cells [3]. These application policies are orthogonal to those considered by our approach, and could be incorporated for more expressive policies.

Roy *et al.* [31] present Airavat, a Hadoop-based MapReduce framework with enhancements for controlling information flow. Airavat runs on SELinux, using its type enforcement for domain isolation. Airavat modifies HDFS to manage its own security labels. Using these, it implements a custom policy based on differential privacy, to minimize leakage of private data during MapReduce computations. In particular, the MapReduce framework (including Airavat) and reducer implementations are trusted subjects. In contrast, in MLS-BTC there are no trusted subjects external to the OS.

3   MLS Architectures Overview

Before discussing a proposed design for an MLS cloud data store, we briefly review architectures for MLS RDBMS and cloud data stores, generally.

MLS Database Architectures.   Several secure architectures have been previously identified for multilevel databases, *i.e.*, the Woods Hole architectures [9]. Of these designs, the kernelized architecture provides the basis for our work.

The reader is directed to existing survey work for an in-depth description of other MLS database architectures, *i.e.*, the trusted subject, integrity lock and distributed architectures [28]. For a *kernelized architecture*, multilevel relations are decomposed into single-level relations managed external to the TCB. Different ways to decompose relations, and different ways of managing the resultant single-level data, lead to variants of the kernelized design. In the Hinke-Schaefer architecture [18], there are no trusted components outside the kernel; other variants include SeaView [13] and Lock Data Views [34]. In all variants, multiple single-level untrusted subjects manage the (decomposed) single-level relations.

MLS Cloud Data Stores. No prior MLS database work applies to data stores with relaxed ACID properties, to databases that do not encode relational models, or to databases lacking fixed schemas. We find mandatory access control (MAC) to be orthogonal to the transactional properties of relational databases, and believe MLS non-relational stores to be a new and interesting domain.

Indeed, certain design patterns for distributed, cloud data stores seem synergistic with architectures for multilevel relational databases. For example, in MLS systems, information flow restrictions require some data and single-level services to be inaccessible to clients based on its level; data store designs that accommodate partition tolerance and availability in the presence of failures seem to accommodate adaptation to these environments.

Data store designs that employ append-only, log-structured storage managed by the underlying TCB can be implemented using a lock-free design, possibly allowing access to high-readers while a low-update is in progress. Such concurrent access comes at the expense of replacing strong consistency by eventual consistency, which for many stores is part of the intended design. Thus, systems whose data structures use write-ahead logs (*e.g.*, to support a crash-only design [7]) may, in practice, enable eventually-consistent, read-down operations.

For some MLS relational databases, clients access a single database front-end. Data sharding allows a client to independently determine the location of nodes in the cluster holding its data, to contact each node directly. In a replicated architecture—in which trusted front-end agents mediate access to untrusted backend databases [28]—sharding may entirely eliminate the need for trusted front-ends, allowing single-level subjects to interact with the services at their level, directly, to access data at or below their level.

The most common criticism of the Hinke-Schaefer architecture is the difficulty of implementing transactions, due to the need for read-locks and, thus, the possibility of flows that violate MAC policy [15]. Some NoSQL stores, however, sacrifice transactions for scalability, foregoing read locks and, thus, this problem.

4 Kernelized MLS Column-Store

We present the design for an MLS column-store following a kernelized architecture. We call this the *MLS-aware BigTable Clone* (MLS-BTC) design, as it is largely applicable to any data store following the published design of Google's BigTable [8]. To describe MLS-BTC, we adopt basic terminology employed by

Apache HBase, a popular open-source BigTable clone. In our design, *all* policy enforcement is performed by an underlying trusted operating system; it mediates access to all resources and enforces an MLS policy. Single-level clients interact with MLS-aware, single-level applications running on each server, which in turn may access resources using interfaces exposed by the trusted OS. A benefit of this approach is that applications are not involved in MLS policy enforcement and, thus, reside outside the TCB and do not need to be engineered to be trustworthy.

4.1   MLS-BTC Design

In MLS-BTC, each table holds timestamped data, organized by rows and columns, and grouped by column families. Table data is partitioned into *regions* of contiguous rows. A *region server* (RS) manages a set of regions, handling all operations on its assigned regions, and splitting regions that have grown above the configured region size. For persistence, each region is stored to a distributed file system which is, itself, an MLS-aware service following a kernelized architecture, *i.e.*, the underlying trusted OS enforces the policy for accessing stored objects.

Each MLS-BTC node holds multiple untrusted RS instances, one per level. Each RS stores its data to a directory associated with its level, using the MLS-aware distributed file system. There are no explicit labels in the MLS-BTC columnar data. Rather, following the Hinke-Schaefer design, data is stored to labeled operating system objects. This approach is known to be inefficient when database access patterns require data to be labeled at a fine-granularity [17]. In our design, object labels are coarse (per-level tables), the table namespace is partitioned per level, and the table's constituent objects are stored to different per-level file system directories. RS instances access table data at lower levels by explicitly reading from the appropriate per-level directory.

A dedicated per-level master is responsible for management of RS instances at its level. These duties include table creation, load-balancing regions across RS instances, and handling RS failures. As master instances require knowledge of tables at lower levels, each master is MLS-aware. The master instances and RS instances coordinate through a distributed locking/synchronization system.

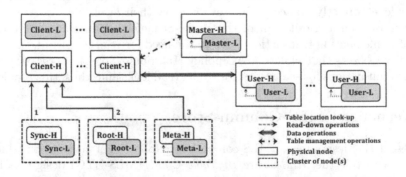

Fig. 1. MLS-BTC Component Relationship

MLS-BTC follows a BigTable architecture, using three special classes of region servers: the *Root* RS, *Meta* RS, and *User* RS. These servers help a client locate the RS hosting a specific table row, using the same region lookup mechanism used in BigTable (Fig. 1). A client locates the Root RS for its level via the distributed synchronization service at its session level. Next, the client contacts the Root RS to find the appropriate Meta RS for the request. The client contacts the Meta RS to find the location of the User RS managing the region for the requested row. Finally, the client contacts the User RS to access the row data.

Concept of Operations. In a typical MLS system, a user session must be associated with a sensitivity level, used to limit access to resources in accordance with MLS policy. For the MLS-BTC system, the sensitivity level of a user's session is associated, statically, with the level of the network interface on which the request is received[1]. A client communicates with the per-level region servers to manipulate table data (*e.g.*, get, put, delete, multi-row scan). A single-level client can write to tables at its session level, and read tables at or below its session level. A client communicates with a per-level master server to issue certain administrative functions (*e.g.*, create, list, or delete tables, add to or drop from column families). The *list-tables* operation returns data for any tables at or below the client's session level.

Design Features. The primary design goals of the MLS-BTC system are: (a) to defer all MLS policy enforcement to the underlying TCB; (b) to use no trusted subjects external to the OS, avoiding extending the TCB boundary, *e.g.*, no trusted proxies or trusted front-ends to communicate between processes at different levels; (c) re-use existing code for untrusted subjects, minimizing the modifications required to make these MLS-aware; (d) expose a familiar client API. We ensure all MLS functionality is deferred to the underling OS by re-designing MLS-BTC components as untrusted subjects following a Hinke-Schaefer design. Re-using existing code for untrusted subjects with only small modification allows us to leverage complex, feature-rich server behavior, and future upgrades to that code, without extending the TCB boundary. A familiar API—such as one compatible with an existing, column-oriented store—will allow MLS-BTC to support legacy applications and simplify new application development.

4.2 Prototype Implementation

Each node in an MLS-BTC cluster is a platform running a trustworthy operating system enforcing an MLS policy. The prototype currently implements this component using SELinux, configured to enforce a MAC policy based on the Bell-LaPadula confidentiality model [16]. The prototype's untrusted subjects are based on a number of existing open-source components, running either unmodified or with small modification:

---

[1] We admit labeling interfaces imposes some deployment inflexibility, adopting it for simplicity; in Sect. 4.2 (Limitations), we suggest more flexible and complex designs.

- *MLS-aware Master & Region Servers.* The prototype re-uses the Master
  Server and Region Server (RS) components of HBase [14], modified to be
  MLS-aware HBase (MA-HBase) components. This provides clients with a
  cross-domain read-down capability, constrained by the SELinux MAC policy.
- *MLS-aware Distributed File System.* The prototype re-uses components of
  the Hadoop Distributed File System [33] (HDFS), modified to produce an
  MLS-aware HDFS (MA-HDFS) component. Details for the design and architecture of the MA-HDFS component are reported in prior work [27].
- *Per-Level Locking/Synchronization Services.* The prototype re-uses components of Zookeeper [19] to implement a distributed synchronization and lock-
  ing service, available at each level. We configure and re-use these components,
  wholesale, as single-level subjects on separate nodes.

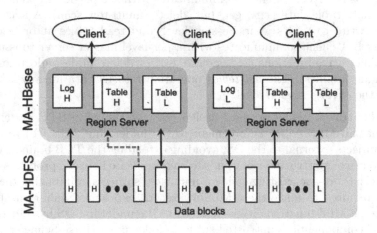

Fig. 2. MLS-BTC Table Storage

Table and Directory Organization. The MA-HBase cluster jointly manages a set of per-level tables (Fig. 2). As in HBase, there are three types of table:
the root table, the meta table, and the user tables; each are maintained by the
Root RS, Meta RS and User RS, respectively. The root and meta table nam-
ing conventions are unchanged. The user table namespace, however, is divided
by level (*level.TableName*), to signal to an RS when a read-down operation is
required to access table data at a lower level.

HBase stores all table data under a configurable root location in HDFS. The
directories under this root (*e.g.*, */hbase*) include a directory tree holding write-
ahead log (HLog) data, and a tree holding per-region table (HFile) data. The
constituent HLog and HFile objects managed by MA-HDFS are stored under a
per-level root location (*e.g.*, */level/hbase*).

For both MA-HBase tables and MA-HDFS directories, the *level* indicator,
used to partition the namespace and invoke read-down logic, is a human-readable
string administratively associated with an SELinux sensitivity level.

Information Flow. The information flow between a client application (local or remote) and an MA-HBase server process is constrained by the system's MAC policy (see Fig. 3). An application can only communicate with an MA-HBase process running at its session level. When the application requires read-access to a table at a lower level, it must request the RS running at its session level to perform a read-down on its behalf. If the application attempts to contact an RS running at some lower level, the underlying trusted OS prohibits this.

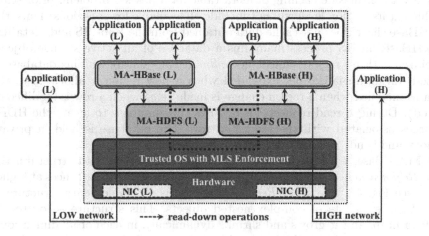

Fig. 3. Information Flow in MLS-BTC

The information flow between an MA-HBase RS process and an MA-HDFS server process is similarly contained: An MA-HBase RS may only communicate with MA-HDFS processes running at the same level. Thus, RS requests to read HLog or HFile data at lower levels must be issued from the RS to an MA-HDFS process at the same level, which in turn performs the read-down operations.

Read-Down Requests. Next, we explain how MLS-BTC handles write, read and read-down requests. This involves two steps: locating the appropriate User RS by the client, and handling the request by the User RS. In MA-HBase, we distinguish between two types of RS: the *authoritative* RS and the *surrogate* RS. The authoritative RS is the "owner" of an allocated region. It runs at the sensitivity level of the corresponding table and updates the MA-HDFS files for storing the row data associated with its regions. The surrogate RS runs at the client's session level and is responsible for handling read-down requests for table data managed by an authoritative RS at a lower level. This is required since a client cannot communicate directly with any lower-level authoritative RS instance. The number of authoritative and surrogate RS instances running on a node is defined administratively through MA-HBase configuration files.

When a client requests access to a row in a table at its session level, it locates the authoritative User RS associated with the row, via the Root and Meta RS.

When the client requests access to a row in a table at a lower level, *i.e.*, a read-down operation, the process is slightly different. The Meta RS recognizes the difference between the client's session level and table's level, and responds to the client with the location of an appropriate surrogate RS, rather than the authoritative RS. In turn, when the client contacts the surrogate RS with a request to read a row at a lower level, the surrogate RS performs a read-down operation to the MA-HDFS resources storing the row data. Since the Meta RS does not read-down on every meta table scan to retrieve region information at lower levels, client-side caching of meta table data poses a problem: prior scans of tables at its session level will not include all available lower regions. Thus, the MA-HBase client does not cache data obtained from the Root RS and Meta RS.

In HBase, an RS process maintains a database of all active storage objects associated with its region, called the *onlineRegions* database. This database is created during initialization, expanded when a new region is allocated to the RS, and modified when a region change is made (*e.g.*, when a row is modified or deleted). During a read request, an RS uses this database to locate the HDFS resources associated with the requested row. The database is held in private memory and is not visible to other RS processes.

In MA-HBase, each authoritative RS maintains a new data structure, the *onlineRegionsCache*, to expose its region data to surrogate RS instances at higher levels (see Fig. 4). The *onlineRegions* database is a complex data structure: a map of maps of lists of complex nested objects. This structure is located in the Java heap, and it grows and shrinks dynamically, in each of its dimensions. To expose its contents to higher levels, some form of IPC must be employed. Using shared memory (*i.e.*, re-implementing it as a library using the Java Native Interface) would be non-trivial. For example, POSIX shared memory sizes are defined at creation time, limiting the dynamic growth of the structure. Further, such a library would require new, custom logic for memory management; the lack of coordination between the memory managers—*i.e.*, Java's garbage collection and the management of the shared memory pool under the native library—would be especially problematic. Instead, the data is exposed using file-based IPC.

On a write request, the RS services the request, updates the *onlineRegions* database, flushes all recent modifications to MA-HDFS, then serializes the in-memory *onlineRegions* database to an *onlineRegionsCache* file. This file is stored to a RAM disk, accessible to surrogate RS instances at higher levels. The surrogate RS can read-down to the RAM disk, to de-serialize and interpret the data structure in response to read-down requests. Using the *onlineRegionsCache* database, a surrogate RS locates the MA-HDFS objects associated with table data at lower levels, and requests these from MA-HDFS processes running at the surrogate's level. Serializing the *onlineRegions* database required developing a custom serialization class, as standard Java object serialization mechanisms could not be used: the database contains inner classes with non-serializable attributes. Concurrent access to the *onlineRegionsCache* by multiple processes is synchronized using a lock-free, read-and-retry consistency mechanism.

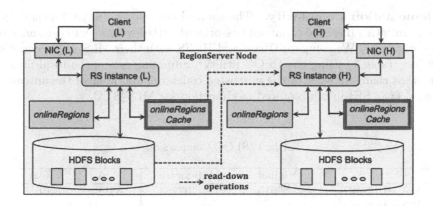

Fig. 4. Authoritative and Surrogate RS Detail

Limitations. The current MLS-BTC prototype system has a number of practical limitations, stemming from our objective to develop a functional, proof-of-concept, non-relational data store that closely follows a kernelized architecture. We summarize those limitations here.

Scalability. The current prototype accommodates policies with a relatively small number of sensitivity levels. For simplicity, the client's session level is associated with a level assigned to the receiving NIC; thus, the number of levels available for the system's policy is constrained by the number of NICs supported by the underlying trusted platform. To support more complex lattice structures, *i.e.*, the "gazillion problem" in MLS design [20], MLS-BTC could be extended with custom trusted components to associate a remote client with a session level and to start services dynamically on behalf of those subjects. The MYSEA system uses such an approach to implement its multilevel LAN concept [22].

Caching. Serializing objects to shared memory and maintaining a consistent image of in-memory objects accessible to subjects at higher levels comes with a performance penalty, discussed further in Sect. 4.3

Surrogate RS. To locate surrogate RS instances, each Meta RS uses a static look-up table providing the authoritative-to-surrogate mappings. For a large MLS-BTC cluster, a runtime mechanism for constructing and managing this mapping should be introduced, allowing authoritative and surrogate servers to enter and leave the system, dynamically.

Meta RS. The current prototype requires all Meta RS instances be co-located, to facilitate read-down to lower meta tables. For a large MLS-BTC cluster, this cannot be guaranteed: a low Master may elect to assign a low meta region to alternate nodes to load-balance. The Master at higher levels could recognize this, and migrate high meta tables in response. This workaround, however, poses a problem when Masters at incomparable levels migrate meta regions to different nodes, forcing higher-level Masters to make an irreconcilable choice regarding with whom they co-locate.

Implementation Complexity. The source lines-of-code (SLOC) metric provides an intuitive measure commonly associated with development cost and software complexity. We compare Hbase and HDFS with their MLS-aware counterparts (see Table 1) using the CLOC utility[2]. Summing across the total lines of source that changed, ~3% of the untrusted codebase, and none of the untrusted codebase (*i.e.*, SELinux), required modification for MLS-BTC.

Table 1. SLOC Comparison

	Original	MLS-aware	Δ	% Δ
MA-HDFS [27]	89615	92263	3314	3.70%
MA-HBase				
Master	8624	8736	116	1.35%
RS	17170	18829	1715	9.99%
Client	7184	7420	270	3.76%
Other	66313	66732	411	0.62%
Total	188906	193980	5826	3.08%

Compatibility. To determine that our prototype data store is functionally compatible with legacy web-applications, while constraining these according to MLS policy, we tested three applications: Titan, Storm and AppScale.

Titan. Titan[3] is an open-source, distributed graph database designed for storing and querying large-scale graphs. We configured Titan to use our prototype as its storage backend. Titan's Gremlin tool was able to manipulate sample graphs stored in the data store: read/write graphs held in tables at the client's level, and read graphs held in tables at lower levels.

Storm. Storm[4] is an open-source, distributed stream processing platform. Storm does not run on Hadoop; however, using an HBase connector[5], Storm can be configured to use HBase as a storage back-end. We configured Storm to store processed data in a table at the client's level. Theoretically, other applications could read this Storm-processed data, either at or below their level.

AppScale. AppScale[6] is an open-source re-implementation of Google's App Engine platform. AppScale supports HBase as a storage back-end, to store a variety of persistent data used by the platform (*e.g.*, user-uploaded content, system metadata). We modified AppScale (v1.7.0) to use our prototype as its primary datastore, rather than the precompiled HBase distributed with AppScale. A sample GAE application, the *guestbook* program, was used to test AppScale's use

---

[2] Count Lines of Code, http://cloc.sourceforge.net
[3] Titan, https://thinkaurelius.github.io/titan/
[4] Storm, https://storm.incubator.apache.org/
[5] https://github.com/jrkinley/storm-hbase
[6] AppScale, http://www.appscale.com/

of the HBase API. The program could read and write to the HBase tables containing user messages at the client's level. Theoretically, the *guestbook* program could be modified to perform explicit read-downs to other table data.

4.3   Prototype Evaluation

To measure the performance of the MLS-BTC prototype, we used the Performance Evaluation (PE) benchmark distributed with HBase and the Yahoo! Cloud Serving Benchmark (YCSB) suite [11].

The PE benchmark implements the same tests used to evaluate performance for BigTable: a sequential read test (Seq-R), random read test (Rand-R), scan test (Scan-R), sequential write test (Seq-W) and random write test (Rand-W); see Chang *et al.* [8] for details. The benchmark employs a hard-coded table name in its tests; we added an option to specify the table to use, for testing read-down.

YCSB is a benchmark framework for evaluating the performance and elasticity of cloud storage systems, and has been employed to benchmark systems like Cassandra, HBase and PNUTS. YCSB provides a set of test workloads, to evaluate different aspects of a system's performance. All six workloads use a similar set of records as test data. For details on the test workloads, see Cooper *et al.* [10]. We followed the recommended test order (A, B, C, F, D, E), which keeps a consistent store size. Test data were loaded prior to running YCSB-A. Before starting YCSB-E, all tables from previous workloads are removed and new test data loaded. In YCSB, all workloads require writes before or during each run; thus, no read-down operation was tested.

Each test in the PE and YCSB benchmarks is executed in one of three scenarios: using 100,000 rows (100K), 500,000 rows (500K) and 1 million rows (1M) workload sizes. Before each trial, all stored HDFS/HBase data are removed, to start each trial from a comparable initial state.

Test Environment.   The test environment consists of twelve nodes evenly distributed across four server blades in one of two racks. Each node is a virtual machine hosted on VMware ESXi 5.0.0. One rack contains three server blades (each, a Dell PowerEdge R710, with 8 CPUs x 2.925 GHz with hyper-threading active, 48GB RAM and Gigabit Ethernet). The other rack holds the remaining server blade (a Dell PowerEdge R610, with 8 CPUs x 2.26 GHz with hyper-threading active, 24GB RAM and Gigabit Ethernet).

Results.   Benchmark results for the MLS-BTC prototype are summarized in Fig. 5. We normalize each trial by the average HBase performance—*i.e.*, mean of three trials under same test conditions with HBase—to obtain an "overhead factor," a positive multiplicand expressing performance relative to HBase. In general, all tests experience performance degradation, which is expected. The degree of degradation, however, varies significantly, impacted by the size and mixture of the workload.

The PE write-tests show substantial degradation, even for the relatively small 100K-row workload. In contrast, the PE read-tests exhibit overheads associated with both table creation and read performance. For most YCSB workloads, the

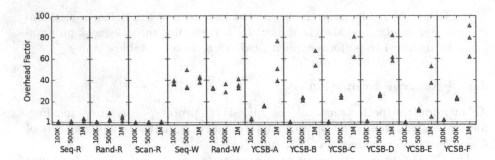

Fig. 5. Prototype performance, normalized by HBase performance

prototype is more than 50× slower than HBase. In all write workloads, performance degradation is the result of both caching the *onlineRegions* database (anytime a row is created or modified) and caching the HDFS namespace (anytime an HBase object stored in HDFS is created or modified). During read-down, performance degradation is the result of reading the cached *onlineRegions* database to handle each read-down request (see Fig. 6). During other reads, the degradation is attributable to lack of client-side caching of server metadata. In general, the most significant performance bottlenecks are associated with the caching of the data structures that maintain the HDFS namespace (FSImage) on the name node and the location of HDFS blocks (BlockMap) on the data nodes [27].

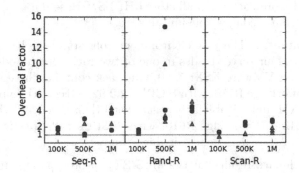

Fig. 6. Highlight of Fig. 5, including read-down performance (blue circle)

We note that more data is required for a rigorous characterization of system performance, but the observed data suggests a rough order-of-magnitude degradation: writes are processed ~40× slower than HBase; reads are ~1.5–8× slower, and mixed read/write workloads can experience ~10–90× slowdown. In Sect. 5, we discuss some more general outcomes and lessons learned.

5   Discussion

We find the general approach of caching objects for Hinke-Schaefer is not appropriate for large, distributed systems in Java. In particular, the lack of efficient IPC mechanisms for object sharing calls into question the viability of the kernelized approach for managing MLS-aware Java applications. Most methods for IPC in Java are bi-directional (*i.e.*, socket-based IPC) or limited by small buffer sizes (*i.e.*, I/O-stream based IPC). As Java lacks an API for shared memory, MLS-BTC re-uses file-based IPC for OS-enforced data sharing. We find this is inefficient for sharing complex, in-memory objects across levels. We discuss some alternatives to get more acceptable performance, based on this observation.

The kernelized approach could be explored after enhancing Java with more flexible OS-enforced IPC interfaces. For example, Kaffe [4] is a research JVM with a process-level abstraction, allowing different Java processes to communicate via shared memory in a controlled way. Supporting HDFS/HBase on Kaffe under SELinux may be promising for kernelized MLS designs with Java-based systems, although Kaffe appears to no longer be maintained.

In the extreme, our experiences could be interpreted as evidence that the Hinke-Schaefer approach should be abandoned, and others explored. For example, using the trusted front-end variant of the kernelized architecture, a carefully-engineered trusted proxy may significantly improve performance. In MLS-BTC, such a trusted proxy can forward client requests at different levels to the appropriate RS processes, eliminating the need to cache the *onlineRegions* data structure. The challenge is to design a small, covert channel-free subject whose responsiveness and efficiency removes significant bottlenecks; this is challenging given that "responsive" and "channel-free" tend to be mutually exclusive.

In the integrity lock architecture—in which an untrusted DBMS stores all multilevel objects [28]—cryptographic protection on objects prevent untrusted subjects from processing aggregate queries. This requires the trusted front-end to be more complex, to perform extra post-query processing. Many key-value models, however, do not support an API with queries returning aggregate objects: simple put, get, delete semantics return individual objects. The integrity lock architecture can be re-explored in this context, re-evaluating all prior criticisms. In particular, the architecture's (known) signaling channel could be bounded with respect to some popular datastore API.

6   Conclusion

We have presented a design for an MLS-aware column-store, faithfully following the kernelized design pattern. The resulting system, MLS-BTC, constrains access to resources at different levels, enabling read-down without any trusted subjects outside the TCB. Our prototype evaluation questions the practicality of the kernelized design approach to manage MLS-aware, untrusted Java-based applications. MLS-BTC is a distributed system based on HBase using SELinux for MLS policy enforcement; it is the first cloud-scale data store following a high-assurance design.

References

1. Anderson, J.: Computer security technology planning study. Technical Report ESD-TR-73-51, The Mitre Corporation, Air Force Electronic Systems Division, Hanscom AFB, Bradford, MA (October 1972)
2. Apache Accumulo Project. Apache Accumulo user manual version 1.5 (2014)
3. Apache HBase Project. The Apache HBase reference guide (2014)
4. Back, G., Hsieh, W.C.: The KaffeOS java runtime system. ACM Trans. Program. Lang. Syst. 27(4), 583–630 (2005)
5. Bates, A., Mood, B., Pletcher, J., Pruse, H., Valafar, M., Butler, K.: Detecting co-residency with active traffic analysis techniques. In: Proc. of the ACM Workshop on Cloud Computing Security, pp. 1–12 (2012)
6. Buxbaum, P.: Clouds at the edge: Army intel program deploys first tactical cloud computing node in Afghanistan. Geospatial Intelligence Forum 11(2), 8–12 (2013)
7. Candea, G., Fox, A.: Crash-only software. In: USENIX Workshop on Hot Topics in Operating Systems, pp. 67–72 (2003)
8. Chang, F., Dean, J., Ghemawat, S., Hsieh, W.C., Wallach, D.A., Burrows, M., Chandra, T., Fikes, A., Gruber, R.E.: Bigtable: A distributed storage system for structured data. ACM Trans. Comput. Syst. 26(2), 4:1–4:26 (2008)
9. Committee on Multilevel Data Management Security. Multilevel data management security. Technical report, Air Force Studies Board (1983)
10. Cooper, B.: YCSB core workloads (2010), http://goo.gl/NJBV4L
11. Cooper, B.F., Silberstein, A., Tam, E., Ramakrishnan, R., Sears, R.: Benchmarking cloud serving systems with YCSB. In: Proc. of the ACM Symp. on Cloud Computing, pp. 143–154 (2010)
12. Currie, W., Seddon, J.J.: A cross-country study of cloud computing policy and regulation in healthcare. In: Proc. of the 22nd European Conf. on Information Systems (2014)
13. Denning, D.E., Lunt, T.F., Schell, R.R., Shockley, W.R., Heckman, M.: The SeaView security model. In: Proc. of the IEEE Symp. on Security and Privacy, pp. 218–233 (1988)
14. George, L.: HBase: The Definitive Guide. O'Reilly Media (2011)
15. Graubart, R.D.: A comparison of three secure DBMS architectures. In: Database Security III: Status and Prospects, pp. 167–190 (1989)
16. Hanson, C.: SELinux and MLS: Putting the pieces together. In: Proc. of the Annual SELinux Symp. (2006)
17. Hinke, T.: Secure database management system architectural analysis. In: 2nd Aerospace Computer Security Conf., pp. 65–72 (1986)
18. Hinke, T.H., Schaefer, M.: Secure data management system. Technical Report RADC-TR-75-266, System Development Corp. (November 1975)
19. Hunt, P., Konar, M., Junqueira, F., Reed, B.: Zookeeper: Wait-free coordination for internet-scale systems. In: Proc. of the USENIX Annual Technical Conf. (2010)
20. Irvine, C.: A multilevel file system for high assurance. In: Proc. of the 1995 IEEE Symp. on Security and Privacy, pp. 78–87 (May 1995)
21. Irvine, C.E., Acheson, T., Thompson, M.F.: Building trust into a multilevel file system. In: Proc. 13th National Computer Security Conf., pp. 450–459 (1990)
22. Irvine, C.E., Nguyen, T.D., Shifflett, D.J., Levin, T.E., Khosalim, J., Prince, C., Clark, P.C., Gondree, M.: MYSEA: The Monterey security architecture. In: Proc. of the ACM Workshop on Scalable Trusted Computing, pp. 39–48 (2009)
23. Jaeger, T.: Operating System Security. Morgan and Claypool Publishers (2008)

24. Killion, T.: Future naval capabilities. In: NDIA 15th Annual Science and Engineering Technology Conf. (April 9, 2014)
25. Konkel, F.: Intelligence community builds cloud infrastructure. In: FCW (September 2013), http://goo.gl/mfYjV9
26. McDermott, J., Montrose, B., Li, M., Kirby, J., Kang, M.: Separation virtual machine monitors. In: Proc. of the Annual Computer Security Applications Conf., pp. 419–428 (2012)
27. Nguyen, T., Gondree, M., Khosalim, J., Irvine, C.: Towards a cross-domain MapReduce framework. In: IEEE MILCOM 2013, pp. 1436–1441 (2013)
28. Notargiacomo, L.: Architectures for MLS database management systems. In: Information Security: An Integrated Collection of Essays, pp. 439–459 (1995)
29. Porche III, I.R., Wilson, B., Johnson, E.-E., Tierney, S., Saltzman, E.: Data flood: Helping the Navy Address the Rising Tide of Sensor Information. Rand (2014)
30. Ristenpart, T., Tromer, E., Shacham, H., Savage, S.: Hey, you, get off of my cloud: Exploring information leakage in third-party compute clouds. In: Proc. of 16th ACM Conf. on Computer and Communications Security, pp. 199–212 (2009)
31. Roy, I., Setty, S.T.V., Kilzer, A., Shmatikov, V., Witchel, E.: Airavat: Security and privacy for MapReduce. In: Proc. of the USENIX Conf. on Networked Systems Design and Implementation (NSDI), p. 20 (2010)
32. Shockley, W., Schell, R., Thompson, M.F.: The importance of high assurance computers for command, control, communications, and intelligence systems. In: Aerospace Computer Security Applications Conf., pp. 331–342 (December 1988)
33. Shvachko, K., Kuang, H., Radia, S., Chansler, R.: The hadoop distributed file system. In: Proc. of the 26th IEEE Symp. on Mass Storage Systems and Technologies (MSST), pp. 1–10 (2010)
34. Stachour, P.D., Thuraisingham, B.: Design of LDV: A multilevel secure relational database management system. IEEE Trans. Knowledge and Data Engineering 2, 190–209 (1990)
35. Stonebraker, M., Cetintemel, U.: One size fits all: an idea whose time has come and gone. In: Proc. of the Intl. Conf. on Data Engineering, pp. 2–11 (2005)
36. Watson, P.: A multi-level security model for partitioning workflows over federated clouds. In: Proc. of the IEEE Conf. on Cloud Computing Technology and Science (CloudCom), pp. 180–188 (November 2011)
37. Weissman, C.D., Bobrowski, S.: The design of the force.com multitenant internet application development platform. In: Proc. of the 2009 ACM SIGMOD Conf., pp. 889–896 (2009)
38. Wu, R., Ahn, G.-J., Hu, H., Singhal, M.: Information flow control in cloud computing. In: Proc. of the Conf. on Collaborative Computing (CollaborateCom), pp. 1–7 (October 2010)
39. Xu, Y., Bailey, M., Jahanian, F., Joshi, K., Hiltunen, M., Schlichting, R.: An exploration of L2 cache covert channels in virtualized environments. In: Proc. of the ACM Workshop on Cloud Computing Security, pp. 29–40 (2011)
40. Zhang, Y., Juels, A., Reiter, M.K., Ristenpart, T.: Cross-VM side channels and their use to extract private keys. In: Proc. of the ACM Conf. on Computer and Communications Security, pp. 305–316 (2012)

Idea: Optimising Multi-Cloud Deployments with Security Controls as Constraints

Philippe Massonet[2], Jesus Luna[1], Alain Pannetrat[1], and Ruben Trapero[3]

[1] Cloud Security Alliance (Europe), United Kingdom
{jluna,apannetrat}@cloudsecurityalliance.org
[2] Centre d'Excellence en Technologies de l'Information et de la Communication, Belgium
philippe.massonet@cetic.be
[3] Department of Computer Science, Technische Universitat Darmstadt, Germany
rtrapero@deeds.informatik.tu-darmstadt.de

Abstract. The increasing number of cloud service providers (CSP) is creating opportunities for multi-cloud deployments, where components are deployed across different CSP, instead of within a single CSP. Selecting the right set of CSP for a deployment then becomes a key step in the deployment process. This paper argues that deployment should take security into account when selecting CSP. This paper makes two contributions in this direction. First the paper describes how industrial standard security control frameworks may be integrated into the deployment process to select CSP that provide sufficient levels of security. It also argues that ability to monitor CSP security should also be considered. The paper then describes how security requirements may be modelled as constraints on deployment objectives to find optimal deployment plans. The importance of using cloud security standards as a basis for reasoning on required and provided security features is discussed.

Keywords: Cloud, Security, Deployment, Optimization, Security controls, Security service level agreements, Monitoring.

1 Introduction

The growing number of CSP offering infrastructure services (IaaS) opens up opportunities for benefitting from the advantages of deploying applications over multiple CSP. Multi-cloud systems (MCS) [8, 9] involve deploying components of a single application on more than one CSP. There are multiple reasons justifying MCS and they range from improving fault tolerance, to minimizing cost of deployment, or to improving response time by deploying components closer to customer locations.

However moving to MCS raises new challenges such as being able to deploy, undeploy and redeploy easily from one CSP to another. The general approach taken in this paper is to build a deployment model that is independent of any specific CSP [7]. A deployment process then transforms the CSP independent deployment model into an executable deployment plan for a specific CSP. Such a model must capture functional as well as non-functional deployment requirements. In this paper we suggest to

F. Piessens et al. (Eds.): ESSoS 2015, LNCS 8978, pp. 102–110, 2015.

take security into account by expressing security requirements in terms of cloud security control frameworks such as CCM [1]. Such frameworks provide some degree of security assurance and transparency, because auditors evaluate the security of a cloud service against a set of "controls" chosen from a reference "security control framework". Security controls remain high level and can be implemented in many different ways.

Workgroups at the European Network and Information Security Agency (ENISA) and the National Institute of Standards and Technology (NIST) have identified [3, 4] that specifying security service level objectives (SLO) in security Service Level Agreements (secSLA) is a useful tool to establish common semantics supporting the description of security assurances. In order to present the key elements driving the adoption of useful SLO to select cloud providers, this paper explores (i) how security controls and security level objectives can be modelled as constraints on a cloud deployment, and (ii) how these security constraints can be used to select the best set of cloud providers on which to deploy the different components of a MCS. The paper discusses the use of SLO as a basis for continuous monitoring of cloud service security as future work at the end of the paper. The arguments presented in this paper are the result of our research and field experience in relevant academic/industrial projects (e.g., EU funded SPECS [5], CUMULUS [6] and PAASAGE [7]), standardization bodies (e.g., NIST, ETSI and ISO/IEC), and related Cloud Security Alliance workgroups (e.g., Cloud Trust Protocol –CTP- and Service Level Agreements).

This paper is organized as follows: Section 2 describes the concept of multi-cloud and identifies challenges for deployment. Section 3 argues that industry standards for security controls and security level objectives should be used to allow comparing side-by-side cloud providers. Section 4 describes how selecting the best set of cloud providers for a cloud deployment can be modelled as a combinatorial optimization problem, and that security controls and SLO can be modelled as constraints. The section concludes with some preliminary experimental results.

2   Multi-Cloud Applications and Case Study

Figure 1 illustrates a MCS where a cloud application is deployed both in private and public clouds in different countries and jurisdictions. The objective is to locate the web server close to customers in order to reduce response time as much as possible on mobile devices such as smartphones and tablets. The database server (DBS) is kept in a private cloud in Spain at the head office of the company while the web servers (WS) and application servers (AS) are deployed in public clouds located close to customers in the United Kingdom and Germany.

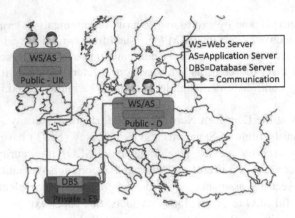

Fig. 1. A multi-cloud deployment in Europe

In this example, it is realistic to suppose that each one of the CSPs involved has implemented its own security mechanisms, controls and policies. Under these circumstances the customer might be also faced with the following security assurance challenges: How to select the right public CSP, based on the component's security requirements? How to continuously monitor the overall cloud infrastructure to assess that all security requirements are fulfilled? Because each provider in the MCS can implement their security controls in a different way, what is the aggregated/overall security assurance level provided to the customer?

The provision of security assurance to the cloud customer in the presented MCS scenario covers many challenges such as cloud provider selection, continuous security monitoring or aggregation of security levels. This paper focuses on the first challenge i.e., selecting the right set of CSP's so that security requirements are satisfied when deploying the application.

3   Model-Based Deployment of Multi-Cloud Applications

This section introduces the context of this paper, namely that deployment should be based on models, and that deployment models should be optimised with respect to objectives and requirements.

3.1   Model-Based Deployment Workflow

This section briefly describes the different phases of the deployment workflow and components [7]. The figure below shows three workflow phases: configuration, deployment and execution. The configuration phase involves building a model of the application components to be deployed. This involves describing in a model the artifacts to be deployed, the communication links between artifacts, the scalability requirements of each artifact, … The model includes a description of security requirements for each deployable artefact. In the deployment phase the deployment model is analysed by a component called the "Reasoner" to produce an optimal deployment plan that meets deployment objective and constraints defined in the model. In terms of security the "Reasoner" component will compare and match security requirements to CSP features. In the execution phase of the workflow the "Adapter", "Deployer" and "Execution engine" execute the deployment plan resulting in a multi cloud deployment that is monitored by the "Adapter". The "Adapter" controls the run-time feedback loop by analysing the monitoring data and performs run-time adaptations. From the security point of view monitoring data about security controls is analysed. If model violations cannot be solved at run-time, then control is passed to a design-time feedback loop where the MCS is stopped and the "Reasoner" calculates a new deployment plan that solves the model violations.

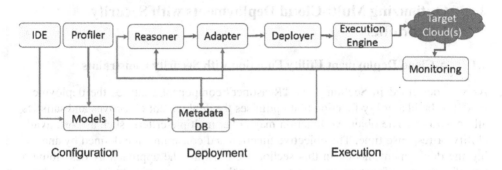

The rest of the paper focuses on the utility function that is used by the "Reasoner" to find an optimal deployment plan that includes security constraints.

3.2   Modelling Deployment Security Requirements

The MCS is assumed to be composed of several components, and it is assumed that each component can be deployed separately on cloud resources. Figure 2 shows a fragment of the deployment meta-model that shows how security meta-concepts are related to deployable artifacts. "Components" are deployable artifacts and can "require" a "security control" in order to be deployed. "Security control" can be "provided" by "Cloud Providers". "Security Controls" are abstract and are difficult to monitor. "SLO" on the other hand are measurable and are related to "SecurityControls".

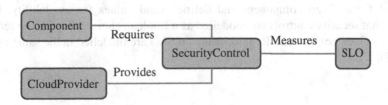

Fig. 2. Security Control Concept in the Meta-model

From the security perspective these meta-concepts provide the basis for matching required "security controls" and "SLO" with CSP "security controls" and "SLO". In order to produce an executable deployment plan, a cloud provider must be selected for each application component. To address the security assurance and transparency issues discussed earlier in the paper, most CSPs would submit their service to certification by independent third party auditors, based on well-established standards such as ISO/IEC 27001 [10], PCI-DSS [11], or CCM [1] for example. Security control frameworks can be complemented with security Service Level Agreements (secSLA). This approach is based on the assessment of measurable SLOs in secSLAs.

4   Optimizing Multi-Cloud Deployments with Security Constraints

4.1   A Cloud Deployment Utility Function with Security Constraints

As was described in Section 3 the "Reasoner" component analyses the deployment model to build a utility function that optimises the deployment objectives and satisfies all constraints. The objective function may cover multiple criteria such as cost, availability or response time. The objective function and constraints are defined by analysing the deployment model. In this section we illustrate the approach by describing a specific objective function and constraints for the running case study. The utility function is then used to produce a deployment plan that makes trade-offs between security and other constraints.

The objective function shown in line (1) of the table below minimises the total deployment cost, i.e. the sum of the costs of all the application components. In the process it must assign a cloud provider to each component. Line (2) defines the deployment cost of a component as the sum of cost for the virtual machines and the cost of the storage measured in terms of I/O operations. The function "provider(c)" in line (5) returns the cloud provider that has been selected for a given component. Line (3) shows that the application to be deployed is composed of three components: two application/web servers, and one database server. It also defines "P" the list of 5 potential cloud providers for deploying the components. Providers 0 and 1 are private clouds, and the others are public clouds. Line (4) defines some bounds for the total cost of a single deployment. Line (6) shows how an availability constraint may be defined for a given component and defines valid values for availability. Line (7) shows that security controls are modelled as a Boolean choice: they can be required or not for a given component. Requirements on SLO are modelled in the same manner in line (8).

$$(1) \; Min \sum_{comp=1}^{comp\_max} comp\_deployment\_cost_{comp}$$

(2) comp_deployment_cost(c) =
$requiredVM(c) * vm\_cost(provider(c)) + requiredIO(c) *$
$IO\_cost(provider(c)))$

(3) c ∈ C, C={ws/as, ws/as, dbs}, P={0, ..., 4}

(4) $comp\_deployment\_cost \in (minCost, maxCost)$, (5) $provider(c) \in P$

(6) $requiredAvailability(c) \geq a$, $where$ 99,9 ≤ a ≤ 99,9999

(7) $requiredSC(c, sc) = b, where \; b \in \{0,1\}, and \; sc \in SecurityControls$

Fig. 3. Equations for minimising deployment cost and selecting a provider per component

To illustrate how security is modelled Figure 4 shows a partial decomposition of CCM security control AIS-04 into several intermediate security controls, e.g. "Prevent improper disclosure", and finally into three secSLOs that can be effectively assessed and monitored in a cloud infra-structure (provided they have been documented using a model like NIST [8]). Take for example "Country level anchoring" SLO which is defined as follows: "this attribute indicates that all processing operations applicable to the resource only take place within a set of predefined countries". The value associated with such SLO is "a vector of strings representing a two-letter ISO-3166-1 country code". This SLO allows expressing a constraint on the jurisdiction in which a cloud deployment can be made, and to subsequently monitor that the deployment has not moved outside of this jurisdiction.

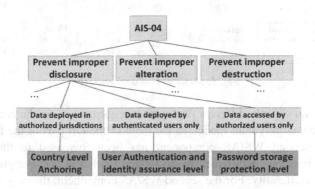

Fig. 4. Decomposition of control AIS-04 into SLOs

To illustrate how the constraints on SC and SLO are instantiated consider the following constraints " $(a) \; requiredSC(ws/as, AIS04) = 1$, (b) requiredSLO(ws/as, CountryLevelAnchoring, DE)=1". The first constraint requires that the provider that is selected for deploying the "ws/as" component must have implemented the

"AIS-04" security control. "AIS-04" describes general mechanisms for controlling data exchange between jurisdictions. If we want to limit data exchange to a list of specific jurisdictions, then we need to add constraint (b) that requires providers to support SLO "CountyLevelAnchoring" and limit the location of data to Germany ("DE").

4.2   Preliminary Experimental Results

The above optimisation problem was modelled with a constraint programming solver [12]. This section describes some preliminary experimental results, by showing different security requirements and the corresponding deployment models. In Figure 5 we see a first deployment model produced by the constraint program. The right part of the figure shows the cloud provider attributes. The deployment model that has been produced by the constraint program shows the deployment that minimises total cost of the deployment. As can be seen in Figure 5, provider 1 that is located in Spain has been selected for all three components because it is the least expensive provider that satisfies all constraints. The total deployment cost is 8014 euro per month. In fact all cloud providers satisfy all component constraints except provider 2, because he does not provide the security controls that are required by the second WS/AS component.

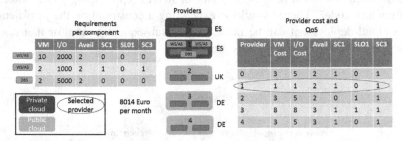

Fig. 5. Minimal cost deployment

Figure 6 shows slightly different deployment requirements on the components and the resulting new deployment model. In the table of the left part of the figure, availability for the second WS/AS component has been increased to three nine, i.e. "99,999". The previous deployment is no longer a solution because provider 1 only offers "99,99" availability. For the second WS/AS component, the only two providers that offer "99,999" availability are Providers 3 and 4 both located in Germany. Provider 4 has been selected because it is less expensive than provider 3. The figure shows the resulting deployment where the second WS/AS component is deployed on provider 4. The other two components are deployed as before on provider 1.

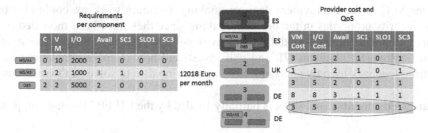

Fig. 6. Deployment for higher availability

Figure 7 shows another change in the deployment requirements and the resulting new deployment model. An SLO, e.g. "requiredSLO(ws/as, CountryLevelAnchoring, DE)=1" is now required for the second WS/AS component. This is shown in red in the table of the left part of the figure. Provider 4 is no longer a solution because it does not offer the required SLO. Even though it is located in Germany it does not provide any data location monitoring data. Provider 3 is the only provider that offers this SLO and is thus the selected provider even though it is the most expensive.

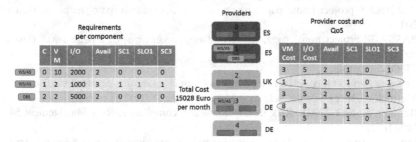

Fig. 7. Deployment with extra security

5   Discussion and Conclusions

Selecting could providers for MCS components is a special case of the general problem of workload placement. The main contribution of this paper is to show that security requirements can deploying workloads to multiple cloud providers using industry standard SC and SLO such as CCM [1]. Related efforts for selecting cloud providers [14] have focused on comparing cloud providers by measuring the level of security they provide. Compared to this paper this work focuses on security and does not take into account other types of requirements. Furthermore it requires that every security requirement be measurable. In our approach we do not attempt to measure security levels, since requirements are expressed as constraints. This approach is more flexible when it is difficult to quantity the requirements. In future work security requirements could be integrated into the objective function provided that they can be quantified. Other related research efforts have worked on integrating security into SLAs by describing fine grained security properties in a security property specification language [15]. In this paper we have combined coarse grained SC requirements with finer grainer SLO in the deployment decision making. In future work we could also integrate security properties in the decision making since they have been modelled in the security deployment meta-model. Future work will integrate the utility function into a cloud deployment platform [7], and will investigate how to make adaptations in the run-time feedback loop by analysing SLO monitoring data to solve SLO violations.

Acknowledgements. This work is partially funded by the EU FP7 PaaSage project.

References

1. Cloud Control Matrix (2011),
   http://www.cloudsecurityalliance.org/cm.html
2. Cloud Security Alliance. The Security, Trust & Assurance Registry (STAR),
   https://cloudsecurityalliance.org/star/ (last access: 2014)
3. Dekker, M., Hogben, G.: Survey and analysis of security parameters in cloud SLAs across the European public sector (2011), http://www.enisa.europa.eu/
4. NIST, Cloud Computing: Cloud Service Metrics Description (RATAX) (2014)
5. SPECS home page, http://specs-project.eu/ (last access: 2014)
6. CUMULUS project home page, http://www.cumulus-project.eu (last access: 2014)
7. PASSAGE project home page, http://www.passage-project.eu/ (last access: 2014)
8. Cloud computing,
   http://en.wikipedia.org/wiki/Cloud_computing#Multicloud
9. Multi cloud, http://en.wikipedia.org/wiki/Multicloud
10. Brenner, J.: ISO 27001: Risk management and compliance. Risk Management 54(1), 24 (2007)
11. Industry, Payment Card. Data security standard. Requirements and Security Assessment Procedures, Version 3 (2013)
12. Choco Solver, http://www.emn.fr/z-info/choco-solver/
13. NIST, Cloud Computing: Cloud Service Metrics Description (RATAX). Working document (2014)
14. Garcia, J.L., Vateva-Gurova, T., Suri, N., Rak, M., Liccardo, L.: Negotiating and Brokering Cloud Resources based on Security Level Agreements. In: CLOSER, pp. 533–541. SciTePress (2013)
15. Pannetrat, A., Hogben, G., et al.: D2.1 Security-aware SLA specification language and Cloud security dependency model., CUMULUS project deliverable (2013)

Idea: Towards an Inverted Cloud

Raoul Strackx[1], Pieter Philippaerts[2], and Frédéric Vogels[2]

[1] iMinds-DistriNet, University of Leuven, Leuven, Belgium
firstname.lastname@cs.kuleuven.be
[2] University College Leuven-Limburg, Leuven & Limburg, Belgium
firstname.lastname@ucll.be

Abstract. In this paper we propose the concept of an inverted cloud infrastructure. The traditional view of a cloud is turned upside down: instead of having services or infrastructure offered by a single provider, the same can be achieved by an aggregation of a multitude of *mini providers*. Even though the contribution of an individual mini provider in an inverted cloud can be limited, the combination would nevertheless be significant. We propose an architecture for an implementation of an inverted cloud infrastructure to allow mini providers to offer processor time. Security and efficiency can be achieved by building upon Intel's new SGX technology.

Keywords: protected module architectures, cloud computing, secure execution.

1 Introduction

Cloud computing is a relatively young trend, having gained much traction in the past few years. Businesses and individuals alike rely on cloud computing for a variety of tasks. Cloud computing introduces a certain dynamicity to the way we work with computers. For example, a company might want to ensure availability of its systems during peak periods. One way of handling this situation consists of simply leasing additional virtual servers whenever necessary. This both reduces the total cost of ownership and allows us to deal with unexpected traffic surges robustly.

While cloud computing offers the financial advantage of sharing hardware costs among users, a significant investment is unfortunately still required from cloud providers. Meanwhile, the devices used to access cloud services are themselves underutilized. Estimates of the average utilization of computer processors vary between 5% and 15% [19].

This paper proposes a new approach, called the *inverted cloud*, which allows these idle resources to be harvested. The traditional view of a cloud is turned upside down: Instead of having services offered by a single provider, the same is achieved by an aggregation of *mini providers*. Even though the contribution of an individual mini provider in an inverted cloud can be limited, the combination would nevertheless be significant. To concretize the idea, this paper works out an architecture for an inverted cloud system to share processor time.

F. Piessens et al. (Eds.): ESSoS 2015, LNCS 8978, pp. 111–118, 2015.

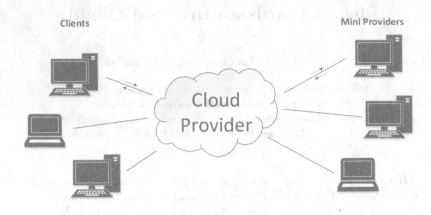

Fig. 1. An overview of the inverted cloud architecture

The next section investigates the requirements of the proposed architecture. Section 3 works out the architecture in some detail, building upon Intel's new SGX technology to ensure security and efficiency. Section 4 discusses work related to our proposal, and finally Section 5 concludes the paper.

2   Overview and Requirements

In a standard cloud computing setting, clients send small requests to a cloud provider where they are processed and their result returned. Due to page constraints we take a more abstract approach where clients provide work packages that need to be executed. In Section 5 we discuss briefly how tasks could be split in work packages. Note that work packages are not required to be completely self-contained; They may still access other resources on the network.

In an inverted cloud setting the processing of these work packages is outsourced to mini providers. This reduces the task of a cloud provider to tracking mini providers entering and leaving the network and balancing work loads. Figure 1 displays an overview of this architecture. Clients and mini providers are only represented as two disjoint sets for clarity; There is no technical reason that a client could not also be a mini provider.

In order to be useful, the system must comply with a number of requirements. The rest of this section gives an overview of the requirements that were considered during the design of the communication and execution protocol.

Requirement 1: Secrecy of Computation Mini providers will execute work packages on behalf of clients, but they should not be able to tell *what* is being computed. The input data and the algorithm should be kept secret.

Requirement 2: Integrity of Computation Mini providers must attest that the correct, unmodified work package was executed and produced the returned result. Any manipulation of the result must be detectable by the cloud provider.

Requirement 3: Secure Execution The intermediate or output data must not be leaked; The computation must take place in a protected environment and the output should be protected.

Requirement 4: Performance The design of the system must allow for an efficient implementation.

3   Architecture

In order to execute work packages in full isolation on the resources of a mini provider, many attack vectors need to be addressed. We first introduce recent advances in protected-module architectures and sandboxing that enable such strong security guarantees, before discussing how these security primitives can be combined in a novel way to build an inverted cloud.

3.1   Protected-Module Architectures

Providing strong isolation of code and data on commodity computing devices is challenging. Operating systems have grown too complex to be able to guarantee that no defects exist that could compromise this isolation.

Security measures have been proposed to significantly raise the bar for attackers [5, 24, 18, 11, 27], but vulnerabilities (e.g., buffer overflows [13, 25]) in commodity applications and operating systems continue to be exploited on a daily basis. Any system relying on such a huge trusted computing base (TCB) cannot offer the security guarantees required to build a large inverted cloud.

Recent years many research projects have taken an alternative approach. Instead of relying on a huge TCB where the operating system provides all possible required services to applications, *protected-module architectures* (PMAs) have been proposed that provide only a minimal set of security primitives, which can be implemented in hardware [12, 7, 14, 9, 23] or by a very limited-sized hypervisor [22, 26, 8]. The exact set of primitives offered depends on the proposed security architecture, but all provide strong isolation of modules: modules are in complete control of their own memory space. Any attempt to access their memory region by code executing outside the module at *any* privilege level (including from other modules) will be blocked. Modules can only be accessed through an interface that they expose explicitly. Hence, even when the system is infested with malware, secrecy and integrity of protected modules remain guaranteed.[1] Recent work by Agten et al. [2, 1] and Patrignani et al. [17, 16] proves that high-level software properties can also be guaranteed at low-level by relying on PMA's memory protection and inserting proper checks at compile time.

PMAs avoid the snowball effect of ever-growing TCBs by using the operating system's services, while *not* trusting them completely. Strackx et al. [22] implement an example where an SSL-connection is set up between a protected

---

[1] In practice modules may be manipulated before protection is enabled. Such attacks are detected when the correct execution of modules is *attested* to a remote verifier or when modules attempt to access previously stored secrets.

module and a remote server. By placing application and SSL logic within protected modules, only encrypted network packets cross the modules' protection boundaries. While the operating system's services are still relied upon (e.g. for network access), these services need not be trusted. Even though kernel-level malware may modify, replay or drop network packets, confidentiality and integrity of data exchanged is guaranteed. This effectively reduces the power of an in-kernel attacker to that of a network-level attacker.

In 2013, Intel announced Software Guard eXtensions (SGX), a PMA to be implemented in their processors in the near future. Intel SGX provides even stronger security guarantees than most state-of-the-art research architectures. It not only protects modules (called "enclaves" in SGX terminology) against software-level attacks, it also guards against hardware-level attacks by ensuring that enclaves are stored unencrypted only within the processor package.

3.2 Isolating Enclaves

Protected-module architectures provide strong isolation of data stored. To enable easy integration of protected modules in legacy software, PMAs such as Intel SGX execute modules in the same address space as the rest of the application. As a result, inputs to modules do not need to be marshalled but modules can simply be provided with pointers to unprotected memory areas. Unfortunately, this also enables malicious modules to extract data stored in the same address space, or, even more worrysome, to attack the operating system.

Avonds et al. [4] and Strackx et al. [20] propose a mechanism to isolate potential attack vectors in an application (e.g., parsers) from likely attack targets (e.g., cryptographic keys). Using an approach similar to PMAs, they divide large applications in multiple compartments where each compartment can only be accessed through the interface they expose explicitly. Access to the operating system is also heavily restricted: compartments can at initialization time indicate which system calls will never be issued. Once a system call has been disabled by a compartment, it can never be re-enabled. An application that is properly compartmentalized will disable all system calls from likely attack vectors and an attacker will need to compromise multiple compartments before reaching attack targets.

For our inverted cloud infrastructure, we will use a similar sandboxing technique to protect mini providers from potentially malicious work packages. Before execution, work packages are placed in a compartment without any system call privileges. Only a very limited interface is provided to return execution results to the cloud provider.

3.3 Executing Opaque Workloads

In order to guarantee correct and safe execution, our protocol operates in two phases (see Figure 2). First, at initialization, a container enclave C is deployed on the mini provider and a public-private key pair is generated. The mini provider is now part of the inverted cloud and can receive work packages for execution. In the second phase the private key is used by the cloud provider to send encrypted work packages to the mini provider. These work packages will be passed to the container enclave C where they are decrypted and executed in complete isolation.

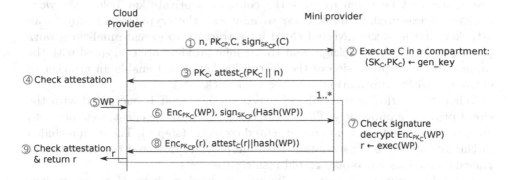

Fig. 2. The communication protocol to share work packages

Phase 1: Initialization. We assume that the mini provider supports the safe execution of potentially malicious work packages. Such support can be implemented as a stand-alone application that provides compartmentalization, or it can be integrated in an existing application such as a browser.

When the mini provider contacts the cloud provider (CP) to take part in the inverted cloud, she is provided with a (signed) container C and a nonce n (step 1 in Figure 2). After the signature of C has been verified, C is placed in a compartment. Adhering to the principle of least privilege, the compartment should only provide support to connect to the cloud provider. All system calls should be disabled.

In step 2, container C is passed to a function in the compartment where it is loaded in an enclave. Next, a private-public key pair is generated within enclave C's protection boundaries and sealed to its identity, ensuring that the cryptographic keys can be used by future instances of C. The public key is returned together with an attestation[2] guaranteeing that C was executed correctly (step 3). The enclosed nonce n ensures freshness. After the cloud provider verified the attestation (step 4) and determined that PK_C was generated by a correctly deployed and unmodified container C, the cryptographic key is stored. Work packages send to the mini provider will always be encrypted using this public key.

Phase 2: Putting Mini Providers to Work. When a client sends a work package to the cloud provider, a mini provider is selected and the work package is encrypted with the mini provider's public key.[3] As this key is only accessible from container C, attackers intercepting an encrypted work package WP in step

---

[2] An attestation is a log of inputs and outputs signed by a trusted entity (e.g., the mini provider's platform). To avoid linkability of attestations, Intel SGX uses a more privacy-friendly attestation scheme [3]. For clarity, we simply state what is attested.

[3] The client could encrypt the work package with the mini-provider's key when the cloud provider is not trusted, at the cost of increased communication overhead.

6 cannot decrypt it. A signature is also provided to guarantee that the work package originates from the cloud provider.

In step 7, the mini provider checks the signature and decrypts the work package. Care must be taken to *erase* the container's private key before the work package is executed. Failure to do so may leak the cryptographic key to an attacker that uses the inverted cloud infrastructure to execute malicious work packages. As all work packages sent to the mini provider are encrypted with the same public key, possession of the decryption key would enable an attacker to extract sensitive information from the work packages.

When the work package finished executing, its result is encrypted with the cloud provider's public key. This result is sent back to the client, together with an attestation that the container executed correctly (step 8). The hash included in the attestation log ensures that the mini provider executed the correct work package and does not replay an old result.

The mini provider can now receive its next work package. However, enclave container C needs to be destroyed and recreated to regain access to the sealed SK_C and PK_C keys. Enabling access to the sealed cryptographic keys on disk poses a security vulnerability as it may be exploited by a malicious work package.

An alternative solution would be to load work packages in their own enclave in encrypted form. The decryption key could be passed in encrypted form to container C that provides it to the enclave after its correct set up was verified using local attestation [3]. Small bootcode at the beginning of the work package would enable decryption of the rest of the package.

4   Related Work

Outsourcing work packages to other devices is not a new idea, but related work can either not provide strong security guarantees or incurs significant overhead. The SETI@Home project, for example, distributes work loads to analyze radio signals in search for extraterrestrial life. To defend against malicious nodes returning incorrect results, the same work load is send to two different nodes. Only when both results match, the result is accepted. While effective, this approach wastes computing power and cannot guarantee confidentiality of work loads.

Recently Miller et al. [10] proposed a modification to Bitcoin. While Bitcoin clients are only required to execute otherwise useless computational workloads, they propose Permacoin, a protocol where clients are also required to store large, arbitrary volumes of data. As data can be easily confidential and integrity protected, clients can be used to store sensitive data. We propose a complementary protocol enabling (potentially malicious) mini providers to operate on that data.

Parno et al. [15] take an alternative approach to "instill greater confidence in computations outsourced to the cloud." Instead of providing a safe execution environment, they execute on encrypted data directly. A proof of correct execution is returned to the requester that requires less computing power than the original computation. A similar approach could be used to invert the cloud, but performance overhead to still too huge to be applied in practice.

The most related work was presented by Dunn et al. [6]. They propose the use of TPM primitives to prevent the analysis of malware. While similar, their approach cannot defend against a powerful hardware attacker. Naturally, isolation of potentially malicious work packages is also not in scope of their work.

5   Conclusion and Future Work

We have presented the concept of an *inverted cloud*. A major advantage of this approach is that cloud providers do not have to invest in resources themselves, but simply allocate resources of so-called *mini providers* to clients.

We have proposed an architecture for an inverted cloud service where clients can buy processing time. We have shown that an implementation of this idea is feasible when taking advantage of the new Intel SGX technology. The proposed architecture takes into account a number of requirements that ensure the secrecy and integrity of the computations.

In future work, we will implement and evaluate our proposed architecture. This will give us more insight in the performance of mini providers and the overhead induced by the network and communication protocol. We expect that our approach can be easily applied to solve computationally intensive work loads that can easily be split in short, parallel tasks. Applications that require long, sequential computation power may be harder to port to an inverted cloud setting, especially when mini providers may unexpectedly disconnect from the cloud network. For such work loads we look into two complementary research directions. First, mini providers may return intermediate results in the form of new work packages. Computation may then be continued by other mini providers. Second, to reduce the impact of network packet overhead and quickly disconnecting mini providers, we are looking at related work [21] to accompany work packages with digital credits. Mini providers that successfully finish execution of work packages or return intermediate results, are awarded a portion of the credits.

References

1. Agten, P., Jacobs, B., Piessens, F.: Sound modular verification of C code executing in an unverified context. In: Accepted for publication in Proceedings of the Symposium on Principles of Programming Languages (POPL 2015) (2015)
2. Agten, P., Strackx, R., Jacobs, B., Piessens, F.: Secure compilation to modern processors. In: Computer Security Foundations Symposium (2012)
3. Anati, I., Gueron, S., Johnson, S., Scarlata, V.: Innovative technology for CPU based attestation and sealing. In: HASP 2013 (2013)
4. Avonds, N., Strackx, R., Agten, P., Piessens, F.: Salus: Non-hierarchical memory access rights to enforce the principle of least privilege. In: Zia, T., Zomaya, A., Varadharajan, V., Mao, M. (eds.) SecureComm 2013. LNICST, vol. 127, pp. 252–269. Springer, Heidelberg (2013)
5. Cowan, C., Pu, C., Maier, D., Hintony, H., Walpole, J., Bakke, P., Beattie, S., Grier, A., Wagle, P., Zhang, Q.: Stackguard: Automatic adaptive detection and prevention of buffer-overflow attacks. In: USENIX Security Symposium (1998)
6. Dunn, A.M., Hofmann, O.S., Waters, B., Witchel, E.: Cloaking malware with the trusted platform module. In: USENIX Conference on Security (2011)

7. Intel Corporation. Software Guard Extensions Programming Reference (2013)
8. McCune, J.M., Li, Y., Qu, N., Zhou, Z., Datta, A., Gligor, V., Perrig, A.: TrustVisor: Efficient TCB reduction and attestation. In: Security and Privacy (2010)
9. McCune, J.M., Parno, B., Perrig, A., Reiter, M.K., Isozaki, H.: Flicker: An execution infrastructure for TCB minimization. In: EuroSys 2008 (2008)
10. Miller, A., Shi, E., Juels, A., Parno, B., Katz, J.: Permacoin: Repurposing bitcoin work for data preservation. In: Security and Privacy (May)
11. Nikiforakis, N., Piessens, F., Joosen, W.: HeapSentry: Kernel-assisted protection against heap overflows. In: Rieck, K., Stewin, P., Seifert, J.-P. (eds.) DIMVA 2013. LNCS, vol. 7967, pp. 177–196. Springer, Heidelberg (2013)
12. Noorman, J., Agten, P., Daniels, W., Strackx, R., Herrewege, A.V., Huygens, C., Preneel, B., Verbauwhede, I., Piessens, F.: Sancus: Low-cost trustworthy extensible networked devices with a zero-software trusted computing base. In: USENIX Security Symposium (2013)
13. One, A.: Smashing the stack for fun and profit. Phrack Magazine 7(49) (1996)
14. Owusu, E., Guajardo, J., McCune, J., Newsome, J., Perrig, A., Vasudevan, A.: OASIS: on achieving a sanctuary for integrity and secrecy on untrusted platforms. In: Computer & Communications Security (2013)
15. Parno, B., Gentry, C., Howell, J., Raykova, M.: Pinocchio: Nearly practical verifiable computation. In: Security and Privacy (S&P 2013) (2013)
16. Patrignani, M., Agten, P., Strackx, R., Jacobs, B., Clarke, D., Piessens, F.: Secure compilation to protected module architectures. Accepted for Publication in Transactions on Programming Languages and Systems
17. Patrignani, M., Clarke, D., Piessens, F.: Secure Compilation of Object-Oriented Components to Protected Module Architectures. In: Shan, C.-C. (ed.) APLAS 2013. LNCS, vol. 8301, pp. 176–191. Springer, Heidelberg (2013)
18. Philippaerts, P., Younan, Y., Muylle, S., Piessens, F., Lachmund, S., Walter, T.: Code Pointer Masking: Hardening Applications against Code Injection Attacks. In: Detection of Intrusions and Malware, and Vulnerability Assessment
19. Poniatowski, M.: Foundation of Green IT. Prentice Hall (2009)
20. Strackx, R., Agten, P., Avonds, N., Piessens, F.: Salus: Kernel support for secure process compartments. Accepted for publication in Endorsed Transactions on Security and Safety
21. Strackx, R., Lambrigts, N.: Idea: State-continuous transfer of state in protected-module architectures. In: Piessens, F., Caballero, J., Bielova, N. (eds.) ESSoS 2015. LNCS, vol. 8978, pp. 43–50. Springer, Heidelberg (2015)
22. Strackx, R., Piessens, F.: Fides: Selectively hardening software application components against kernel-level or process-level malware. In: CCS (2012)
23. Strackx, R., Piessens, F., Preneel, B.: Efficient Isolation of Trusted Subsystems in Embedded Systems. In: Security and Privacy in Communication Networks (2010)
24. Strackx, R., Younan, Y., Philippaerts, P., Piessens, F.: Efficient and effective buffer overflow protection on ARM processors. In: WISTP 2010 (2010)
25. Strackx, R., Younan, Y., Philippaerts, P., Piessens, F., Lachmund, S., Walter, T.: Breaking the memory secrecy assumption. In: EuroSec 2009 (2009)
26. Vasudevan, A., Chaki, S., Jia, L., McCune, J., Newsome, J., Datta, A.: Design, implementation and verification of an extensible and modular hypervisor framework. In: Security and Privacy (2013)
27. Younan, Y., Philippaerts, P., Cavallaro, L., Sekar, R., Piessens, F., Joosen, W.: PAriCheck: an efficient pointer arithmetic checker for C programs. In: ASIACCS 2010 (2010)

OMEN: Faster Password Guessing
Using an Ordered Markov Enumerator

Markus Dürmuth[1], Fabian Angelstorf[1], Claude Castelluccia[2], Daniele Perito[2],
and Abdelberi Chaabane[2]

[1] Ruhr-University Bochum, Germany
markus.duermuth@rub.de
[2] INRIA, France
{claude.castelluccia,daniele.perito}@inria.fr

Abstract. Passwords are widely used for user authentication, and will likely remain in use in the foreseeable future, despite several weaknesses. One important weakness is that human-generated passwords are far from being random, which makes them susceptible to guessing attacks. Understanding the adversaries capabilities for guessing attacks is a fundamental necessity for estimating their impact and advising countermeasures.

This paper presents OMEN, a new Markov model-based password cracker that extends ideas proposed by Narayanan and Shmatikov (CCS 2005). The main novelty of our tool is that it generates password candidates according to their occurrence probabilities, i.e., it outputs most likely passwords first. As shown by our extensive experiments, OMEN significantly improves guessing speed over existing proposals.

In particular, we compare the performance of OMEN with the Markov mode of John the Ripper, which implements the password indexing function by Narayanan and Shmatikov. OMEN guesses more than 40% of passwords correctly with the first 90 million guesses, while JtR-Markov (for $T = 1$ billion) needs at least eight times as many guesses to reach the same goal, and OMEN guesses more than 80% of passwords correctly at 10 billion guesses, more than all probabilistic password crackers we compared against.

Keywords: Authentication, Password guessing, Markov models.

1 Introduction

Password-based authentication is the most widely used form of user authentication, both online and offline. Despite their weaknesses, passwords will likely remain the predominant form of authentication for the foreseeable future, due to a number of advantages: passwords are highly portable, easy to understand for laypersons, and easy to implement for the operators. In fact, while alternative forms of authentication can replace passwords in specific scenarios, they have not been able, so far, to replace them on a large scale [3].

In this work, we concentrate on offline guessing attacks, in which the attacker can make a number of guesses bounded only by the time and resources she is willing to invest. While such attacks can be improved by increasing the resource with

F. Piessens et al. (Eds.): ESSoS 2015, LNCS 8978, pp. 119–132, 2015.

which an attacker can generate and verify guesses (e.g., by using specialized hardware and large computing resources [9,8]), we concentrate here on techniques to reduce the number of guesses required to crack a password. Hence, our approach reduces the attack time independently of the available resources.

Tools commonly used for password cracking, such as John the Ripper (JtR) in dictionary mode, exploit regularities in the structure of password by applying *mangling rules* to an existing dictionary of words (e.g., by replacing the letter a with @ or by appending a number). This is used to generate new guesses from an existing corpus of data, like a dictionary or a previously leaked password database. Weir et al. [24] demonstrated how to use probabilistic context-free grammars (PCFG) to automatically extract such mangling rules from a corpus of leaked passwords, and Narayanan et al. [16] showed that Markov models, which are known to closely represent natural language, can also be used to guess passwords efficiently. We will demonstrate that, while these attacks already have a good guessing efficiency against passwords, the performance can be substantially improved.

This paper presents OMEN, a new Markov model-based password cracker that generates password candidates according to their occurrence probabilities, i.e., it outputs most likely passwords first. As shown by our extensive experiments, OMEN significantly improves guessing speed over existing proposals.

1.1 Related Work

One of the main problems with passwords is that many users choose *weak* passwords. These passwords typically have a rich structure and thus can be guessed much faster than with brute-force guessing attacks. Best practice mandates that only the hash of a password is stored on the server, not the password, in order to prevent leaking plain-text when the database is compromised.

In this work we consider *offline guessing attacks*, where an attacker has gained access to this hash and tries to recover the password *pwd*. The hash function is frequently designed for the purpose of slowing down guessing attempts [20]. This means that the cracking effort is *strongly dominated by the computation of the hash function* making the cost of generating a new guess relatively small. Therefore, we evaluate all password crackers based on the number of attempts they make to correctly guess passwords.

John the Ripper: John the Ripper (JtR) [17] is one of the most popular password crackers. It proposes different methods to generate passwords. In *dictionary* mode, a dictionary of words is provided as input, and the tool tests each one of them. Users can also specify various mangling rules. Similarly to [6], we discover that for relatively small number of guesses (less than 10^8), JtR in dictionary mode produces best results. In Incremental mode (JtR-inc) [17], JtR tries passwords based on a (modified) 3-gram Markov model.

Password Guessing with Markov Models: Markov models have proven very useful for computer security in general and for password security in particular. They are an effective tool to crack passwords [16], and can likewise be used to accurately estimate the strength of new passwords [5]. Recent independent work [14] compared different forms of probabilistic password models and concluded that Markov models are better suited for estimating password probabilities than probabilistic context-free grammars. The biggest difference to our work is that they only approximate the likelihood of passwords, which does not yield a password guesser which outputs guesses in the correct order, the main contribution of our work.

The underlying idea of Markov models is that adjacent letters in human-generated passwords are not independently chosen, but follow certain regularities (e.g., the 2-gram th is much more likely than tq and the letter e is very likely to follow the 2-gram th). In an n-gram Markov model, one models the probability of the next character in a string based on a prefix of length $n-1$. Hence, for a given string $c_1, \ldots, c_m$, a Markov model estimates its probability as $P(c_1, \ldots, c_m) \approx P(c_1, \ldots, c_{n-1}) \cdot \prod_{i=n}^{m} P(c_i|c_{i-n+1}, \ldots, c_{i-1})$. For password cracking, one basically learns the initial probabilities $P(c_1, \ldots, c_{n-1})$ and the transition probabilities $P(c_n|c_1, \ldots, c_{n-1})$ from real-world data (which should be as close as possible to the distribution we expect in the data that we attack), and then enumerates passwords in order of descending probabilities as estimated by the Markov model. To make this attack efficient, we need to consider a number of details: Limited data makes learning these probabilities challenging (*data sparseness*) and enumerating the passwords in the *optimal order* is challenging.

Probabilistic Grammars-Based Schemes: A scheme based on probabilistic context-free grammars (PCFG) [24] bases on the idea that typical passwords have a certain structure. The likeliness of different structures are extracted from lists of real-world passwords, and these structures are later used to generate password guesses.

Password Strength Estimation: A problem closely related to password guessing is that of *estimating the strength of a password*, which is of central importance for the operator of a site to ensure a certain level of security. In the beginning, password cracking was used to find weak passwords [15]. Since then, much more refined methods have been developed. Typically, so-called pro-active password checkers are used to exclude weak passwords [22,11,1,18,4]. However, most pro-active password checkers use relatively simple rule-sets to determine password strength, which have been shown to be a rather bad indicator of real-world password strength [23,12,5]. The influence of password policies on password strength is studied in [10], and [2] proposes new methods for measuring password strength and applies them to a large corpus or passwords. More recently, Schechter et al. [21] classified password strength by limiting the number of occurrences of a password in the password database. Finally, Markov models have been shown to be a good predictor of password strength while being provably secure [5].

1.2   Paper Organization

In Section 2 we describe the *Ordered Markov ENumerator* (OMEN) and provide several experiments for selecting adequate parameters. Section 3 gives details about OMEN's cracking performance, including a comparison with other password guessers. We conclude the paper with a brief discussion in Section 4.

2   OMEN: An Improved Markov Model Password Cracker

In this section we present our implementation of password enumeration algorithm, enumPwd(), based on Markov models. Our implementation improves previous work based on Markov models by Narayanan et al. [16] and JtR [17]. We then present how OMEN, our new password cracker, uses it in practice.

2.1   An Improved Enumeration Algorithm (enumPwd())

Narayanan et al.'s indexing algorithm [16] has the disadvantage of not outputting passwords in order of decreasing probability. However, guessing passwords in the right order can substantially speed up password guessing (see the example in Section 3). We developed an algorithm, enumPwd(), to enumerate passwords with (approximately) decreasing probabilities.

On a high level, our algorithm discretizes all probabilities into a number of bins, and iterates over all those bins in order of decreasing likelihood. For each bin, it finds all passwords that match the probability associated with this bin and outputs them. More precisely, we first take the logarithm of all n-gram probabilities, and discretize them into levels (denoted η) similarly to Narayanan et al. [16], according to the formula $lvl_i = \text{round}\left(\log(c_1 \cdot prob_i + c_2)\right)$, where c_1 and c_2 are chosen such that the most frequent n-grams get a level of 0 and that n-grams that did not appear in the training are still assigned a small probability. Note that levels are negative, and we adjusted the parameters to get the desired number of levels ($nbLevel$), i.e., the levels can take values $0, -1, \ldots, -(\text{nbLevel}-1)$ where $nbLevel$ is a parameter. The number of levels influences both the accuracy of the algorithm as well as the running time: more levels means better accuracy, but also a longer running time.

For a specific length ℓ and level η, enumPwd(η, ℓ) proceeds as follows[1]:

1. It computes all vectors $\boldsymbol{a} = (a_2, \ldots, a_\ell)$ of length $\ell - 1$, such that each entry a_i is an integer in the range $[0, \text{nbLevel} - 1]$, and the sum of all elements is η. Note that the vectors have $\ell - 1$ elements as, when using 3-grams, we need $\ell - 2$ transition probabilities and 1 initial probability to determine the probability for a string of length ℓ. For example, the probability of the password "password" of size $\ell = 8$ is computed as follows:

$$P(password) = P(pa)P(s|pa)P(s|as)P(w|ss)P(o|sw)P(r|wo)P(d|or).$$

2. For each such vector $\boldsymbol{a}$, it selects all 2-grams $x_1 x_2$ whose probabilities match level a_2. For each of these 2-grams, it iterates over all x_3 such that the 3-gram

---

[1] To ease presentation, we only describe the estimation algorithm for 3-grams. The generalization to n-grams is straightforward.

Algorithm 1 Enumerating passwords for level η and length ℓ (here for $\ell = 4$).[2]

---

function enumPwd(η, ℓ)
  1. for each vector $(a_i)_{2 \leq i \leq \ell}$ with $\sum_i a_i = \eta$
     and for each $x_1 x_2 \in \Sigma^2$ with $L(x_1 x_2) = a_2$
     and for each $x_3 \in \Sigma$ with $L(x_3 \mid x_1 x_2) = a_3$
     and for each $x_4 \in \Sigma$ with $L(x_4 \mid x_2 x_3) = a_4$:
     (a) output $x_1 x_2 x_3 x_4$

$x_1 x_2 x_3$ has level a_3. Next, for each of these 3-grams, it iterates over all x_4 such that the 3-gram $x_2 x_3 x_4$ has level a_4, and so on, until the desired length is reached. In the end, this process outputs a set of candidate passwords of length ℓ and level (or "strength") η.

A more formal description is presented in Algorithm 1. It describes the algorithm for $\ell = 4$. However, the extension to larger ℓ is straightforward.

Example: We illustrate the algorithm with a brief example. For simplicity, we consider passwords of length $\ell = 3$ over a small alphabet $\Sigma = \{a, b\}$, where the initial probabilities have levels

$$L(aa) = 0, \quad L(ab) = -1,$$
$$L(ba) = -1, \quad L(bb) = 0,$$

and transitions have levels

$$L(a|aa) = -1 \quad L(b|aa) = -1$$
$$L(a|ab) = 0 \quad L(b|ab) = -2$$
$$L(a|ba) = -1 \quad L(b|ba) = -1$$
$$L(a|bb) = 0 \quad L(b|bb) = -2.$$

- Starting with level $\eta = 0$ gives the vector $(0, 0)$, which matches to the password **bba** only (the prefix "aa" matches the level 0, but there is no matching transition with level 0).
- Level $\eta = -1$ gives the vector $(-1, 0)$, which yields **aba** (the prefix "ba" has no matching transition for level 0), as well as the vector $(0, -1)$, which yields **aaa** and **aab**).
- Level $\eta = -2$ gives three vectors: $(-2, 0)$ yields no output (because no initial probability matches the level -2), $(-1, -1)$ yields **baa** and **bab**, and $(0, -2)$ yields **bba**.
- and so one for all remaining levels.

2.2   The OMEN Algorithm

As presented previously, the enumeration algorithm, enumPwd(η, ℓ) uses two parameters. These two parameter need to be set properly. The selection of ℓ (i.e.

---

[2] $L(xy)$ and $L(z|xy)$ are the levels of initial and transition probabilities, respectively.

the length of the password to be guessed) is challenging, as the frequency with which a password length appears in the training data is not a good indicator of how often a specific length should be guessed. For example, assume that are as many passwords of length 7 and of length 8, then the success probability of passwords of length 7 is larger as the search-space is smaller. Hence, passwords of length 7 should be guessed first. Therefore, we use an adaptive algorithm that keeps track of the success ratio of each length and schedules more passwords to guess for those lengths that were more effective.

More precisely, our adaptive password scheduling algorithm works as follows:

1. For all n length values of ℓ (we consider lengths from 3 to 20, i.e. $n = 17$), execute enumPwd$(0, \ell)$ and compute the success probability $sp_{\ell,0}$. This probability is computed as the ratio of successfully guessed passwords over the number of generated password guesses of length ℓ.
2. Build a list L of size n, ordered by the success probabilities, where each element is a triple $(sp, level, length)$. (The first element $L[0]$ denotes the element with the largest success probability.)
3. Select the length with the highest success probability, i.e., the first element $L[0] = (sp_0, level_0, length_0)$ and remove it from the list.
4. Run enumPwd$(level_0 - 1, length_0)$, compute the new success probability sp^*, and add the new element $(sp^*, level_0 - 1, length_0)$ to L.
5. Sort L and go to Step 3 until L is empty or enough guesses have been made.

2.3   Selecting Parameters

In this section we discuss several parameters choices and examine the necessary trade-off between accuracy and performance. The three central parameters are: n-gram size, alphabet size and the number of levels for enumerating passwords.

n-**gram size:** The parameter with the greatest impact on accuracy is the size of the n-grams. A larger n generally gives better results as larger n-grams provide a more accurate approximation to the password distribution. However, it implies a larger runtime, as well as larger memory and storage requirements. Note also that the amount of training data is crucial as only a significant amount of data can accurately estimate the parameters (i.e., the initial probabilities and the transition probabilities). We evaluated our algorithm with $n = 2, 3, 4$, results are depicted in Figure 1 (top).

As expected, the larger n is, the better the results are. We have conducted limited experiments with 5-grams, which are not depicted in this graph, but show slightly better results than for 4-grams. As 5-grams require substantially larger running time and memory requirements, for a small performance increase, we decided using $n = 4$.

Alphabet Size: The size of the alphabet is another factor that has the potential to substantially influence the characteristics of the attack. Larger alphabet size means that more parameters need to be estimated and that the runtime and memory requirements increase. In the opposite, a small alphabet size means that not all passwords can be generated.

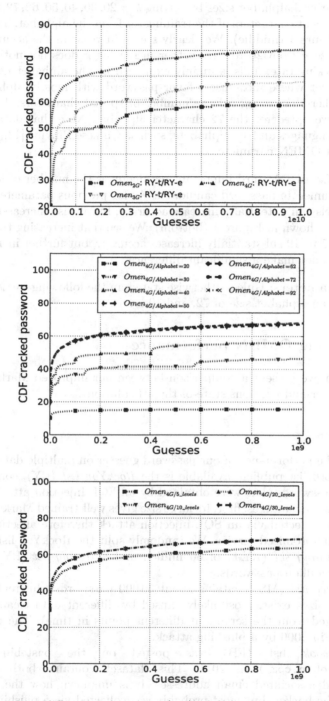

Fig. 1. Comparing different n-gram sizes (top), alphabet sizes (middle), and different number of levels (bottom), for the RockYou dataset

We tested several alphabet sizes by setting $k = 20, 30, 40, 50, 62, 72, 92$, where the k most frequent characters of the training set form the alphabet. The results are given in Figure 1 (middle). We clearly see an increase in the accuracy from an alphabet size k from 20 to 62. Further increasing k does not noticeable increase the cracking rate. This is mainly explained by the alphabet used in the RockYou dataset where most users favor password with mostly alphanumeric characters rather than using a large number of special characters. To be data independent, we opted for the 72 character alphabet. Note that datasets that use different languages and/or alphabets, such as Chinese Pinyins [13], will have to set different OMEN parameters.

Number of Levels: A third important parameter is the number of levels that are used to enumerate password candidates. As for previous parameters, higher number of levels can potentially increase accuracy, but it also increases runtime. The results are shown in Figure 1 (bottom). We see that increasing the number of levels from 5 to 10 substantially increases accuracy, but further increasing to 20 and 30 does not make a significant difference.

Selected Parameters: Unless otherwise stated, in the following we use OMEN with 4-grams, an alphabet size of 72, and 10 levels.

3 Evaluating OMEN performance

In this section, we present a comparison between our improved Markov model password cracker and previous state-of-the-art solutions.

3.1 Datasets

We evaluate the performance of our password guesser on multiple datasets. The largest password list publicly available is the *RockYou list* (RY), consisting of 32.6 million passwords that were obtained by an SQL injection attack in 2009. This list has two advantages: first, its large size gives well-trained Markov models; second, it was collected via an SQL injection attack therefore affecting all the users of the compromised service. We randomly split the RockYou list into two subsets: a *training set* (RY-t) of 30 million and a *testing set* (RY-e) of the remaining 2.6 million passwords.

The *MySpace list* (MS) contains about 50 000 passwords (different versions with different sizes exist, most likely caused by different data cleansing algorithms or leaked from the servers at different points in time). The passwords were obtained in 2006 by a phishing attack.

The *Facebook* list (FB) was posted on the pastebin website (http://pastebin.com/) in 2011. This dataset contains both Facebook passwords and associated email addresses. It is unknown how the data was obtained by the hacker, but most probably was collected via a phishing attack.

Table 1. Summary table indicating the percentage of cracked passwords for 1 billion guesses, or 10 billion when specified

Algorithm	Training Set	#guesses	Testing Set		
			RY-e	MS	FB
Omen	RY-t	10 billion	80.40%	77.06%	66.75%
	RY-t	1 billion	68.7%	64.50%	59.67%
PCFG [24]	RY-t	1 billion	32.63%	51.25%	36.4%
JtR-Markov [16]	RY-t	10 billion	64%	53.19%	61%
	RY-t	1 billion	54.77%	38.57%	49.47%
JtR-Inc	RY-t	10 billion	54%	25.17%	14.8%

Ethical Considerations: Studying databases of leaked password has arguably helped the understanding of users real world password practices and as such, have been used in numerous studies [24,23,5]. Also, these datasets are already available to the public. Nevertheless we treat these lists with the necessary precautions and release aggregated results only that reveal next to no information about the actual passwords (c.f. [7]).

3.2 Comparing OMEN and JtR's Markov Mode

Figure 2 (top) shows the comparison of OMEN and the Markov mode of JtR, which implements the password indexing function by Narayanan et al. [16]. Both models are trained on the RockYou list (RY-t). Then, for JtR-Markov, we fix a target number of guesses T (1 billion or 10 billion), and compute the corresponding level (η) to output T passwords, as required by JtR-Markov.

The curve shows the dramatic improvement in cracking *speed* given by our improved ordering of the password guesses. In fact, JtR-Markov outputs guesses in no particular order which implies that likely passwords can appear "randomly" late in the guesses. This behavior leads to the near-linear curves shown in Figure 2 (top). One may ask whether JtR-Markov would surpass OMEN after the point T; the answer is *no* as the results do not extend linearly beyond the point T; and larger values of T lead to a flatter curve. To demonstrate this claim, we show the same experiment with T equals to 10 billion guesses (instead of 1 billion). Figure 3 shows the results for 1 billion guesses (left) as well as 10 billion guesses (right), and we see that the linear curve becomes *flatter*.

To show the generality of our approach, we compare the cracking performance on three different datasets: RY-e, FB and MS. The ordering advantage allows OMEN to crack more than 40% of passwords (independently of the dataset) in the first 90 million guesses while JtR-Markov cracker needs at least eight times as many guesses to reach the same goal. For the RockYou dataset the results most pronounced: For instance, OMEN cracks 45.2% of RY-e passwords in the first 10 million guesses (see Figure 3 (right)) while JtR-Markov achieves this result after more than 7 billion guesses (for $T = 10$ billion).

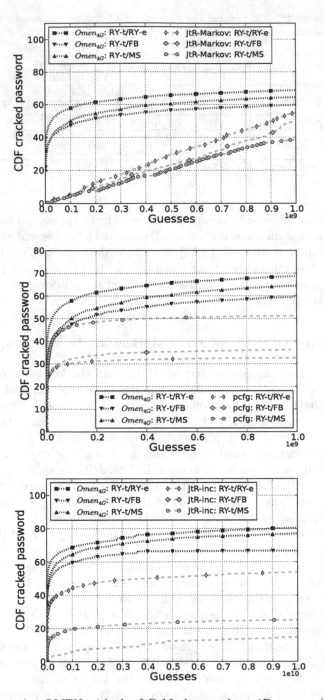

Fig. 2. Comparing OMEN with the JtR Markov mode at 1B guesses (top), with the PCFG guesser (middle), and with JtR incremental mode (bottom)

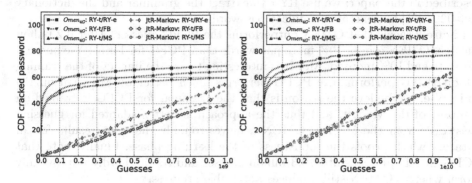

Fig. 3. Comparing OMEN with the JtR Markov mode at 1 billion guesses (left), and at 10 billion guesses (right)

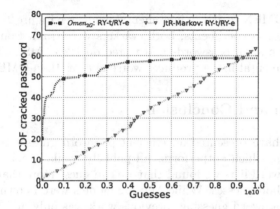

Fig. 4. Comparing OMEN using 2-grams with JtR Markov mode

In the above comparison, OMEN uses 4-grams (c.f. Section 2.3), while JtR-Markov uses 2-grams. To see the effects that this difference has, we provide an additional comparison of OMEN using 2-grams with JtR-Markov, this is given in Figure 4. The results are as expected: JtR-Markov still gives a straight line, which means that OMEN has a better cracking speed. The speed advantage of OMEN can be seen at 1 billion guesses where OMEN cracks 50% of all passwords while JtR-markov cracks less than 10%. At the point T, i.e., when JtR-Markov stops, both algorithms perform roughly the same. Note that since not all parameters (i.e., alphabet size, number of levels etc.) of both models are identical, we have a small difference in the cracking rate at the point T.

3.3 Comparing OMEN and PCFG

Figure 2 (middle) compares OMEN to the PCFG password guesser of Weir et al. [24], based on the code available at [19]. We run it using the configuration as described in the paper: we use RY-t to extract the grammar and the dictionary dict-0294 [25] to generate candidate passwords.

Figure 2 shows that OMEN outperforms the PCFG guesser. After 0.2 billion guesses, OMEN cracks 20% more passwords than PCFG for both RY-e and FB datasets and 10% more for MS. It is interesting to see the impact of the training set on PCFG performance: PCFG performs much better on MS than on FB and RY-e. We believe the reason is that the grammar for PCFG is trained on a subset of the MS list, and thus the approach is better adapted for guessing passwords from the MS list. OMEN achieves roughly the same results for all datasets which proofs the robustness of the learning phase. Finally, note that PCFG mostly plateaus after 0.3 billion guesses and results hardly improve any more, whereas OMEN still produces noticeable progress.

3.4 Comparing OMEN and JtR's Incremental Mode

We also compare OMEN to JtR in incremental mode, see Figure 2 (bottom). Similarly to the previous experiments, both crackers were trained on the RockYou training set of 30 million passwords and tested on RY-e, MS and FB datasets. Clearly, JtR incremental mode produces worse guesses than OMEN.

4 Discussion and Conclusion

In this work, we have presented an efficient password guesser (OMEN) based on Markov models, which outperforms all publicly available password guessers. For common password lists we found that we can guess more than 80% of passwords with 10 billion guesses. While Markov models were known [16] to be an effective tool in password guessing, previous work was only able to output the corresponding guesses in an order dictated by algorithm internals (and pretty much unrelated to their real frequency), OMEN can output guesses in order of (approximate) decreasing frequency and thus dramatically improves real-world guessing speed. Moreover, we performed a number of experiments to assess the impact of different parameters on the accuracy of the algorithm and find optimal parameters. We believe that OMEN can be useful as a preventive measure by organizations to verify that their members do not select "weak" passwords.

References

1. Bishop, M., Klein, D.V.: Improving system security via proactive password checking. Computers & Security 14(3), 233–249 (1995)
2. Bonneau, J.: The science of guessing: analyzing an anonymized corpus of 70 million passwords. In: Proc. IEEE Symposium on Security and Privacy. IEEE (2012)
3. Bonneau, J., Herley, C., van Oorschot, P.C., Stajano, F.: The quest to replace passwords: A framework for comparative evaluation of web authentication schemes. In: Proc. IEEE Symposium on Security and Privacy. IEEE (2012)

4. Burr, W.E., Dodson, D.F., Polk, W.T.: Electronic authentication guideline: NIST special publication 800-63 (2006)

5. Castelluccia, C., Dürmuth, M., Perito, D.: Adaptive password-strength meters from Markov models. In: Proc. Network and Distributed Systems Security Symposium (NDSS). The Internet Society (2012)

6. Dell'Amico, M., Michiardi, P., Roudier, Y.: Password strength: an empirical analysis. In: Proc. 29th conference on Information communications, INFOCOM 2010, pp. 983–991. IEEE Press, Piscataway (2010)

7. Egelman, S., Bonneau, J., Chiasson, S., Dittrich, D., Schechter, S.: It's not stealing if you need it: A panel on the ethics of performing research using public data of illicit origin. In: Blyth, J., Dietrich, S., Camp, L.J. (eds.) FC 2012. LNCS, vol. 7398, pp. 124–132. Springer, Heidelberg (2012)

8. HashCat. OCL HashCat-Plus (2012), http://hashcat.net/oclhashcat-plus/

9. Kedem, G., Ishihara, Y.: Brute force attack on unix passwords with SIMD computer. In: Proc. 8th Conference on USENIX Security Symposium, SSYM 1999, vol. 8. USENIX Association (1999)

10. Kelley, P.G., Komanduri, S., Mazurek, M.L., Shay, R., Vidas, T., Bauer, L., Christin, N., Cranor, L.F., Lopez, J.: Guess again (and again and again): Measuring password strength by simulating password-cracking algorithms. In: Proc. IEEE Symposium on Security and Privacy. IEEE (2012)

11. Klein, D.V.: Foiling the cracker: A survey of, and improvements to, password security. In: Proc. USENIX UNIX Security Workshop (1990)

12. Komanduri, S., Shay, R., Kelley, P.G., Mazurek, M.L., Bauer, L., Christin, N., Cranor, L.F., Egelman, S.: Of passwords and people: Measuring the effect of password-composition policies. In: CHI 2011: Conference on Human Factors in Computing Systems (2011)

13. Li, Z., Han, W., Xu, W.: A large-scale empirical analysis of chinese web passwords. In: Proc. 23rd USENIX Security Symposium, USENIX Security (August 2014)

14. Ma, J., Yang, W., Luo, M., Li, N.: A study of probabilistic password models. In: Proc. IEEE Symposium on Security and Privacy. IEEE Computer Society (2014)

15. Morris, R., Thompson, K.: Password security: a case history. ACM Communications 22(11), 594–597 (1979)

16. Narayanan, A., Shmatikov, V.: Fast dictionary attacks on passwords using time-space tradeoff. In: Proc. 12th ACM conference on Computer and communications security (CCS), pp. 364–372. ACM (2005)

17. OpenWall John the Ripper (2012), http://www.openwall.com/john

18. The password meter, http://www.passwordmeter.com/

19. PCFG Password Cracker implementation Matt Weir (2012), https://sites.google.com/site/reusablesec/Home/password-cracking-tools/probablistic_cracker

20. Provos, N., Mazières, D.: A future-adaptive password scheme. In: Proc. Annual Conference on USENIX Annual Technical Conference, ATEC 1999. USENIX Association (1999)

21. Schechter, S., Herley, C., Mitzenmacher, M.: Popularity is everything: a new approach to protecting passwords from statistical-guessing attacks. In: Proc. 5th USENIX Conference on Hot Topics in Security, pp. 1–8. USENIX Association (2010)

22. Spafford, E.H.: Observing reusable password choices. In: Proc. 3rd Security Symposium, pp. 299–312. USENIX (1992)

23. Weir, M., Aggarwal, S., Collins, M., Stern, H.: Testing metrics for password creation policies by attacking large sets of revealed passwords. In: Proc. 17th ACM Conference on Computer and Communications Security (CCS 2010), pp. 162–175. ACM (2010)
24. Weir, M., Aggarwal, S., de Medeiros, B., Glodek, B.: Password cracking using probabilistic context-free grammars. In: Proc. IEEE Symposium on Security and Privacy, pp. 391–405. IEEE Computer Society (2009)
25. Word list Collection (2012), http://www.outpost9.com/files/WordLists.html

The Heavy Tails of Vulnerability Exploitation*

Luca Allodi

DISI - University of Trento
Via Sommarive 9, Povo, TN, Italy
luca.allodi@unitn.it

Abstract. In this paper we analyse the frequency at which vulnerabilities are exploited in the wild by relying on data collected worldwide by Symantec's sensors. Our analysis comprises 374 exploited vulnerabilities for a total of 75.7 Million recorded attacks spanning three years (2009-2012). We find that for some software as little as 5% of exploited vulnerabilities is responsible for about 95% of the attacks against that platform. This strongly skewed distribution is consistent for all considered software categories, for which a general take-away is that less than 10% of vulnerabilities account for more than 90% of the attacks (with the exception of pre-2009 Java vulnerabilities). Following these findings, we hypothesise vulnerability exploitation may follow a Power Law distribution. Rigorous hypothesis testing results in neither accepting nor rejecting the Power Law Hypothesis, for which further data collection from the security community may be needed. Finally, we present and discuss the *Law of the Work-Averse Attacker* as a possible explanation for the heavy-tailed distributions we find in the data, and present examples of its effects for Apple Quicktime and Microsoft Internet Explorer vulnerabilities.

1 Introduction

Many natural phenomena have been observed to follow heavy-tailed distributions: some notable examples are the frequency distribution of words in a language, the density of metropolitan areas, and the topology of the Internet. Heavy-tailed phenomena significantly differ from 'usual' phenomena that can be easily described by a few point estimations of the distribution. For example, one may consider life-expectancy in a particular country as a quantity that varies only little with respect to the average. In this sense, the average and the standard deviation of the distribution are enough to 'give an idea' of how the distribution

* The author would like to thank Prof. Fabio Massacci at the University of Trento, Julian Williams at the University of Durham (UK) and Matthew Elder at Symantec Corp. for their many useful comments. This project has received funding from the European Union's Seventh Framework Programme for research, technological development and demonstration under grant agreement no 285223 (SECONOMICS). This work is also supported by the Italian PRIN Project TENACE. Our results can be reproduced by utilizing the reference data set WINE-2012-008, archived in the WINE infrastructure.

F. Piessens et al. (Eds.): ESSoS 2015, LNCS 8978, pp. 133–148, 2015.

looks like. For heavy-tailed distributions this does not necessarily hold. For example, if one considers GDP worldwide, the infamous Pareto law kicks in (also known as the *80-20 rule*): 20% of the world population owns 80% of the wealth. In this case, the average income does not provide any real indication of how the distribution looks like, as the top 20% of the population is orders of magnitude richer than the remaining 80%. These distributions are often interesting as they are typically generated by complex phenomena lying behind the observation.

In this paper we provide clear evidence that vulnerability exploitation is described by a heavy-tailed distribution and hypothesise that the distribution may follow a Power Law model. We compare our Power Law hypothesis with two additional candidate models for the data: a Log-Normal hypothesis and an Exponential hypothesis. We proceed by rigorously comparing each generating model against the data, following the methodology described in [8]. We find that the negative exponential distribution hypothesis is ruled out, and that both the power law and the log-normal distribution can be suitable models for the data. These results are in line with those of previous research on malware arrival timings [15], and point toward more research to further investigate the process that generates the observed data.

The results presented in this paper have three main implications:

1. Vulnerability exploitation may be described by laws similar to those followed by natural phenomena (like earthquakes) and self-organizing structures (like cities). In this sense, much in the same way as most earthquakes do not represent a threat for the population, most vulnerabilities may carry negligible risk. This indicates that the classical approach *'I have a vulnerability'* → *'I must fix it'* may be a largely disproportionate reaction to the real threat. An equivalent to this would be to completely evacuate an area typically affected by earthquakes even if the almost totality of earthquakes does not represent a threat for the population.
2. Commonly-used, industry standard definitions of vulnerability risk based on a single number (e.g. scores assigned by security-testing tools) may be incapable of describing the distribution of attacks: a point estimate (a score or an average) is not enough to describe the phenomena and may lead to substantial overspending / misallocation of resources as most events may be orders of magnitude away from the point estimate.
3. A deeper understanding of the attack-generating process may be needed to explain the clear effect we show in the data. In Section 7 we propose the *Law of the Work-Averse Attacker* as a first, informal attempt to explain the heavy-tail effect we observe.

The paper continues as follows: Section 2 introduces the dataset used for the analysis. Section 3 presents *prima facie* evidence of the heavy-tailed distribution of attacks. The paper continues by introducing the models considered for the data (Section 4) and by presenting the methodology and its limitations (Section 5). Results are presented in Section 6. We then discuss this work's implications and present a first attempt to explain the observed effect (Section 7). Related work is discussed in Section 8. Finally, Section 9 concludes the paper.

Table 1. Categories for vulnerability classification and respective number of vulnerabilities and attacks recorded in WINE

Category	Sample of Software names	No. Vulns.	Attacks (Millions)
PLUGIN	Acrobat reader, Flash Player	86	24.75
PROD	Microsoft Office, Eudora	146	3.16
WINDOWS	Windows XP, Vista	87	47.3
BROWSER	Internet Explorer, Firefox	55	0.55
Tot		**374**	**75.76**

2   Data Collection

Symantec runs a data sharing program, the Worldwide Intelligence Network Environment, or WINE in short[1]. The intrusion-prevention telemetry dataset within WINE provides information about network-based attacks detected by Symantec's products. WINE is indexed by *attack signatures IDs*, unique identifiers for an attack detected by the firm's security solutions, which can be linked to the affected CVE, if any, through Symantec's *Security Response*[2] dataset. Further details on the collection process are available in [3]. This experiment's data is referenced and available for sharing at Symantec Research Labs under the WINE Experiment ID *WINE-2012-008*.

We take additional precautions in handling the data to consider for the fact that the prevalence of an attack may depend on the affected software's exposure to attacks. For example, browsers may be mainly exposed to web attacks, while productivity software like MS Outlook may be targeted more often through social engineering and malicious email attachments. We inspected WINE's vulnerabilities and grouped them in eight software categories: Browser, Plugins, Windows, Productivity, Other Operating Systems, Server, Business Software, Development Software. Because WINE consists largely of data from Symantec's consumer security products, we may have a self-selection problem in which certain software categories are not well represented in our sample. We therefore limit our analysis to the first four categories, for which we consider our sample to be representative of exploits in the wild: BROWSER, PLUGIN, WINDOWS and PROD(uctivity). A more detailed discussion on this rationale is given in [5,3]. Our analysis comprises 374 vulnerabilities and 75.7 Million attacks recorded from July 2009 to December 2012. Table 1 reports the identified categories and the number of respective vulnerabilities in WINE.

3   The Heavy Tails of Vulnerability Exploitation

To visualize the heavy tail distribution effect, we report in Figure 1 the histogram distribution of the (logarithmic) attack volumes for each vulnerability

---

[1] https://www.symantec.com/about/profile/universityresearch/sharing.jsp
[2] https://www.symantec.com/security_response/

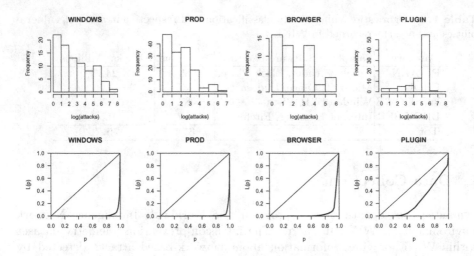

Fig. 1. Top row: histogram distribution of logarithmic exploitation volumes. Bottom row: Lorentz curves for exploitation volumes in the different categories. p % of the vulnerabilities are responsible for L(p)% of the attacks.

in the category (top row) and the respective Lorentz curve distribution (bottom row). The histogram distribution clearly shows that (PLUGIN being an exception we further investigate in Section 6) for WINDOWS, PROD and BROWSER the frequency of vulnerabilities with x attacks is inversely proportional to the logarithm of x. In other words, a (very) small fraction of vulnerabilities is responsible for orders of magnitude more attacks than the remaining vulnerabilities.

A clear way to visualize this is through a Lorentz curve. A Lorentz curve describes the p percentage of the population (of vulnerabilities) that are responsible for the $L(p)$ percent of attacks. The diagonal represents an 'equilibrium state' where each vulnerability is responsible for the same volume of attacks. The further away the two curves are, the higher the 'disparity' in the distribution of attacks per vulnerability. As depicted in Figure 1, for WINDOWS, PROD and BROWSER the two curves are very markedly apart, indicating that the great majority of vulnerabilities are responsible for only a negligible fraction of the risk in the wild. Table 2 reports the distribution of attacks recorded in the wild per vulnerability. We report the top 20, 10 and 5 percent of vulnerabilities and the percentage of attacks in the wild they are responsible for. The most extreme results are obtained for WINDOWS and PROD, for which the top 5% vulnerabilities carry more than 90% of the attacks and the top 10% the almost totality. 'Milder' results are obtained for BROWSER: the top 10% carries 90% of the attacks, but the top 5% carries 'only' 68%, meaning that among the top 10% vulnerabilities attacks are distributed more equally than in other categories. The less extreme result is obtained for PLUGIN, where the distribution of exploitation attempts seems more equally distributed among vulnerabilities.

Table 2. $p\%$ of vulnerabilities responsible for $L(p)\%$ of attacks, reported by software category.

Category	Top $p\%$ vulns.	$L(p)\%$ of attacks
	20%	99.6%
WINDOWS	10%	96.5%
	5%	91.3%
	20%	99.5%
PROD	10%	98.3%
	5%	94.4%
	20%	97.1%
BROWSER	10%	91.3%
	5%	68.2%
	20%	46.9%
PLUGIN	10%	31%
	5%	24%

With this last exception, we observe that a general rule for vulnerability exploitation is that, within any software category, less than 10% of attacked vulnerabilities are responsible for more than 90% of the attacks.

4   Possible Models for the Data

In general, when looking at empirical data it is often difficult to find a perfect fit for a specific distribution. The most cautious way to proceed in this case is to compare different hypotheses against the data. In the heavy-tailed case, models commonly considered as candidates for the data are the *Power Law distribution*, the *Log-Normal distribution*, and the *Exponential distribution* [20].

4.1   Power Law Distribution

A power-law distribution describes a phenomenon whereby the probability of observing an event of size x is proportional to a power of x. Many natural phenomena are known to follow power-law distributions. Earthquakes are a clear example: the probability of observing an earthquake of magnitude x rapidly decreases with the destructiveness of the earthquake[3] [20]. In general, a power law is expressed as:

$$p(x) \sim x^{-\alpha} \tag{1}$$

where x is the measured quantity and α is a *scaling factor* of the distribution. It is easy to see that, if one applies the logarithm on both sides of the equation, one ends up with an equation form of the type $ln(p(x)) = -\alpha ln(x) + c$ which

---

[3] 'Destructiveness' is expressed by measure of the Richter scale which represents the base 10 logarithm of the maximum amplitude of a wave as detected by a seismograph.

is a straight line with (negative) slope α and intercept c. On a log-log plot a distribution following a power law would therefore follow a straight line.

As to the scaling parameter α, most power-law distributions found in Nature are in the range $2 \leq \alpha \leq 3$ [8]. When describing a power law phenomenon the parameter α has some interesting properties attached to it. By calculating the second and third momentum of the normalized power law distribution it is possible to see that depending on the value assumed by α the distribution may have infinite mean ($\alpha < 2$) and infinite variance / standard deviation ($\alpha < 3$). The interested reader can refer to [20] for further details.

In practical terms, a distribution with infinite mean and variance is a distribution that can not be described by point estimates.

4.2   Log-Normal Distribution

Log-normal distributions can be thought as emerging from a *multiplicative effect*. [18] suggests this is for example how one can model biological organisms' growth in weight: as a percentage C of the current weight W_t, such as $W_{t+1} = (C \times W_t) + W_t$. This generates a rapidly growing distribution. If growth in each step of the process is randomly distributed and has finite mean and variance, than because of the central limit theorem one ends up with a normal distribution $N(\sigma, \mu)$ defined in the logarithm of the measure. The function form of a log-normal distribution can therefore be derived from a normal distribution. For further details we refer the reader to [18].

A log-normal distribution has always finite mean and variance, which are therefore more meaningful to consider than in the general power law case.

4.3   Exponential Distribution

An exponential distribution is often used to describe the probability distribution of the distance between independent events that arrive (on average) at a constant rate. A negative exponential is often a less good alternative model to a power law than a log-normal distribution is [8], but we still consider it here for the sake of completeness.

5   Methodology

The central hypothesis around which we build our analysis is:

Hypothesis 1. *Vulnerability exploitation follows a Power Law distribution.*

Following the methodology indicated in [8], we: 1) estimate the parameters for the hypothesised Power Law; 2) Test the suitability of the Power Law model for the data; 3) Compare the Power Law model with alternative possible explanations (i.e. log-normal and exponential)[4].

---

[4] We use the statistical tool R and the PoweRlaw package [26,13]. The scripts are available at https://securitylab.disi.unitn.it/doku.php?id=software

Parameter Estimation. Empirical data is often noisy; in particular when fitting a power law to it, one may find that the data follows a power law only above a certain threshold x_{min}. This is intuitive as in the lower tail small variations in the magnitude of the observation would cause significant noise in the fit. It is generally observed that data points below x_{min} are often better modelled by distributions other than a power law [18]. Exploiting this observation, Clauset et al. [9] suggest to estimate x_{min} by selecting the cutoff that minimizes the distance between the fitted Power Law distribution and the probability distribution of the data. The distance is calculated as the Kolmogorov-Smirnov (KS) statistic, which simply returns the maximum absolute distance between two curves. This way one obtains the x_{min} cutoff that provides the best fit for all $x > x_{min}$. The scaling parameter α is estimated as the parameter that maximizes the likelihood of observing the data given a certain value of α (*maximum likelihood estimation*).

Hypothesis Testing. We now need to estimate how likely the Power Law model is for the data. *Bootstrapping* [11] provides a powerful method to verify the likelihood of the Power Law hypothesis. For each separate data sample DS (e.g. BROWSER) of length n, a *bootstrapped sample* is obtained from the data by randomly choosing with replacement n vulnerabilities from DS. We create 10 thousand bootstrapped samples for each DS. For each bootstrapped sample we then compute the parameter estimation and the relative KS statistic. Then, a *p-value* for the power law hypothesis is obtained by computing the fraction of KS statistics KS' obtained from the sample that are *above* the KS statistic seen from the data. The closest the *p-value* is to the unity, the greatest the evidence for the Power Law Hypothesis[5]. We reject the power-law hypothesis if the resulting p-value is below $p < 0.1$. As noted by Clauset et al., a very good fit ($p > 0.9$) is very unlikely to be found in field data such as ours. We will consider the Power Law model to be *not unreasonable* for $p > 0.1$. This threshold is the same indicated in [8].

Comparison with Other Models. The p-value alone may not be a good-enough indicator of the models' suitability, especially when the data is noisy. Therefore, to more rigorously evaluate the Power Law Hypothesis we compare it with the alternative distributions defined in Section 4.

To compare the models we perform a *log likelihood test*. The idea behind a log likelihood test is to compute the likelihood of observing the data assuming two different originating models: the model with the highest likelihood is, intuitively, the preferred one. A way to see this is to compute the difference in the log likelihoods for the two distributions, R: if R is close to zero, the data has the same likelihood under the two hypotheses; if R is far from zero, the sign of the difference indicates which model is the most suitable for the data. We compute R as $R = log(L(PowerLaw)) - log(L(Alternative))$ where $L()$ is the likelihood function; therefore, a positive sign favours the Power Law hypothesis; a negative sign favours the Alternative.

---

[5] An alternative approach would be to measure the fraction of estimated α' from the bootstrapped sample higher than the α for the original data.

Table 3. Power laws' parameters. The reported α is the median resulting from the bootstrapped process. Significance is reported in bold for $p > 0.1$.

Category	x_{min}	$n_{x \geq x_{min}}$	α	95% Con. In.	p-value
WINDOWS	20	64	1.31	1.22 - 1.64	**0.44**
BROWSER	1010	19	1.60	1.20 - 2.27	**0.52**
PLUGIN	118	80	1.35	1.26 - 2.14	0.00
PROD	267	49	1.50	1.34 - 1.78	**0.84**

The significance of the difference between the two models is given by the size of $|R|$. For values of R close to zero, the sign does not indicate a significant difference between the two models. We use the Vuong test [29] to evaluate the statistical significance of the sign. If the resulting p-value is below 0.1 ($p < 0.1$) we consider the difference to be significant. If not, the two models (Power Law and the Alternative) are effectively indistinguishable with respect to the data.

Limitations. Fitting models to the data asks for as many data points as possible. [8] shows that, indicatively, a distribution with at least 100 points is desirable to make sound conclusions. However, a general estimation of this threshold valid for any distribution is hard to make. Unfortunately exploitation data, especially collected on a significant scale (i.e. worldwide), is difficult to find. No precaution can completely rule out the "overfitting" problem caused by too few data points [8]. In our experiment the worst case is that of BROWSER vulnerabilities, for which we refrain from making any definitive conclusion. For WINDOWS, PLUGIN and PROD we will be slightly bolder. To the best of our knowledge, WINE is the most comprehensive dataset of records of attacks in the wild that is publicly available.

Another limitation is represented by the data collection itself and, indirectly, by the type of software and attacks our results can be considered representative of. The WINE dataset reports mostly attacks recorded against 'consumer platforms', unlikely to receive targeted or 0-days attacks [6]. Our results and conclusions are therefore relevant only for untargeted attack scenarios, and are not representative of 'black swan' attacks for which a dedicated attacker aims at a particular target. This is obvious as a single targeted attack is for us negligible, as it would be in the lower left hand of the Lorentz curves in Figure 1.

6   The Power Law Hypothesis

Parameter Estimation and Hypothesis Testing. Figure 2 reports the log-log plot and a linear fit for the four categories. Attack volumes are reported on the x-axis; the y-axis reports the probability of observing an attack volume equal to x. It is easy to see that for all categories, with the exception of PLUGIN, the data shows a linear trend, which is expected with a Power Law distribution [8,18]. Table 3 reports the model parameters for each category and the p-value for the power law fit. The estimation is done for values $x \geq x_{min}$. This means

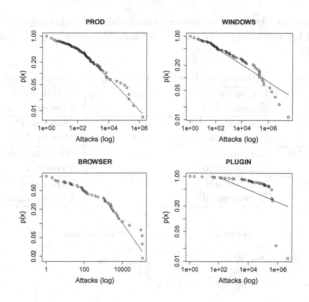

Fig. 2. Log-log plot of vulnerability exploitation by vulnerability rank

that the datasets are further truncated and the estimation is limited to the datapoints left. These are reported in the table under the column $n_{x \geq x_{min}}$. The data points left still allow for an (albeit cautious) discussion. The BROWSER case is critical, as only 19 vulnerabilities are available for the model fitting. For this category the resulting p-value ($p = 0.52$) is higher than the significance threshold of $p > 0.1$ identified by [8], but we refrain from considering this as evidence for the Power Law case. A more significant discussion can be made for the remaining categories. In particular, the Power Law model could be a good candidate to explain the WINDOWS and PROD exploitation distributions. For PLUGIN as a whole, instead, the hypothesis is *ruled out* completely. We will analyse this exception in more detail later in this Section. The α parameter lies in the 1.2-2.2 region for all software categories, indicating a mildly steep to steep curve.

We *do not reject* Hyp. 1 in the cases of WINDOWS, BROWSER and PROD vulnerabilities. We reject Hyp. 1 for PLUGIN.

Comparison with Other Models. In Table 4 we report the results of the log likelihood comparison between Hyp. 1 and the alternative models. A negative sign indicates that the evidence points toward the alternative hypothesis; a positive sign supports the Power Law model. We also report the two-tailed Vuong's significance test; the result is considered significant if the p-value is below 0.1. The log likelihood ratio test for the exponential distribution returns "Not a number" as the fit between the estimated curve and the data is so poor it tends to zero, and the logarithm goes to infinity. The log-normal distribution results slightly favored in the likelihood ratio test for BROWSER and PROD vulnera-

Table 4. Difference in likelihood of alternative models. ∞ indicates a fit so poor that the log likelihood for the alternative goes to infinity. We report significant conclusions for $p \leq 0.1$ in bold.

Category	Alternative	Likelihood difference	Favoured Model
WINDOWS	Log-normal	-1.49	Alternative
	Exponential	∞	**Power Law**
BROWSER	Log-normal	-0.29	Alternative
	Exponential	2.18	**Power Law**
PLUGIN	Log-normal	-5.39	**Alternative**
	Exponential	∞	**Power Law**
PROD	Log-normal	-0.34	Alternative
	Exponential	∞	**Power Law**

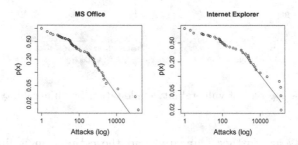

Fig. 3. Log-log plot of volume of exploits in the wild for Microsoft Office (left) and Internet Explorer (right)

bilities, but the small distance from 0 does not make for a solid margin, as the difference may as well be due to sole chance. In general, we find that a log-normal distribution does not perform significantly better than a Power Law in describing our data. For WINDOWS vulnerabilities the evidence is more markedly toward the log-normal distribution, but the difference is again not significant. The case for PLUGIN is, unsurprisingly, sharply in favor of a log-normal distribution. With the exception of PLUGIN, none of the alternative hypothesis in Table 4 provides a better explanation to the data than a Power Law distribution does.

We now narrow down the data analysis to single instances of 'representative' software in each category. We however do not report any more data-fitting results as the fewer and fewer data points would make their interpretation a particularly tricky one.

6.1   Breakdown by Software

Figure 3 reports the distribution of exploitation volume for vulnerabilities affecting Microsoft Office (PROD) and Internet Explorer (BROWSER). For these two software, the log-log plot shows a good linear fit along the data points.

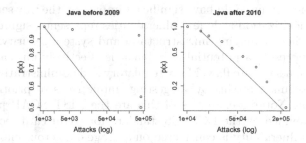

Fig. 4. Log-log plot of volume of exploits in the wild for Java vulnerabilities disclosed before 31-Dec-2009 (left) and after 1-Jan-2010 (rigth)

The numerical results are equivalent to those reported for the respective macro categories. Software in WINDOWS, not reported here for brevity, also confirms the general result. For PLUGIN software, the Power Law fitting is always very low regardless of the considered software. To further investigate this, in Figure 4 we report the distribution for Java vulnerabilities grouped by year of disclosure. The different distribution in attacks for vulnerabilities disclosed before and after 2009 is immediate to see. While for 2009 the Power Law model is clearly a bad fit, for Java vulnerabilities disclosed after 2010 it is supported by the evidence ($p = 0.76$). Neither the log-normal nor the exponential distribution provide a better model for the data. A possible explanation to this temporal effect is that software running in background (such as PLUGIN software generally is) may be seldom updated by users [30]. This may have an influence on the exploitation volumes recorded: looking at the Java case, 2009 is the last year Java was owned by Sun Microsystems, before being acquired by Oracle. This may suggest that pre-2009 vulnerabilities for Sun Microsystem's Java accumulated high exploitation volumes possibly because of users' latency in switching to Orcale's Java. For Java vulnerabilities disclosed after 2010 our results are equivalent to those we obtained for the other categories. This suggests that the heavy-tail effect we observe is present regardless of the software type.

7   Discussion

In this paper we presented evidence that vulnerability exploitation follows a heavy-tailed distribution.

The heavy-tail effect we find is (qualitatively) similar to that shown by the *80-20 Pareto law* of income distribution: the majority of the impact is caused by a small fraction of the population. We showed that, depending on the type of software affected by the vulnerability, as low as 10% of the vulnerabilities may be responsible for more than 90% of the attacks in the wild against that software. The most extreme result is obtained for PROD vulnerabilities, for which 5% of vulnerabilities account for 95% of the attacks.

This observation alone could have significant impact on the way security quantification and prioritization is done. Vulnerabilities represent a significant source of uncertainty when managing infrastructural and system security. Clearly all vulnerabilities represent a potential risk, but it is effectively unclear *how much* risk is attached to a software flaw. Many regulatory and administrative initiatives try to give an estimate of this by suggesting simple rules to prioritize vulnerability treatment. Notable examples of this are the NIST SCAP protocol [25] and guidance provided by the PCI DSS standard for credit card security [10]: a high risk score vulnerability is considered on average dangerous enough to need immediate treatment. This approach has already been questioned in literature [4], and our results point in the same direction: point estimates of vulnerability risk may be widely inappropriate in practice.

7.1   An Explanation Attempt: The Law of the Work-Averse Attacker

We make an attempt at giving an explanation to the possible mechanisms that underlie the heavy-tail effect shown in this paper. We label this the *"Law of the Work-Averse Attacker"*, according to which the average attacker is not interested in procuring and using new reliable exploits if he or she already owns one. The rationale behind this is that once the attacker can attack n systems with one exploit, as long as n is high enough a new reliable (and possibly expensive [2,5]) exploit would not increase n enough to justify the cost (economic or in terms of effort) of deploying a new attack. The effect of this is that attackers focus their efforts in attacking a limited set of vulnerabilities for which reliable exploits exist and are available (for example in the black markets [5]). As a consequence, only a handful of vulnerabilities are consistently attacked over time, and this may generate the heavy-tailed effect shown in this paper. In general,

$$\exists v_{0,t_0}, n : P(v_0, t_0, n) \approx 1 \to \forall v_i \neq v_0 \ P(v_i, t_0, n) \approx 0 \qquad (2)$$

where $P(v, t, n)$ is the probability that an attack against the vulnerability v is successful at time t_0 against n systems, with $n \gg 1$.

If this holds, by looking at exploitation trends in time we would expect that:

Hypothesis 2. *Exploits alternate in 'popularity', i.e. a new one appears only when an old one decays.*

Hypothesis 3. *No two exploits are at the same level of exploitation at the same moment.*

In Figure 5 we report as an example the trends of exploitation of vulnerabilities disclosed in 2010 for Internet Explorer and QuickTime. We plot on a logarithmic scale the volume of attacks against each vulnerability, represented by a distinct line. A simple illustrative example is that of Quicktime, for which it is visible how the emergence of the second exploit follows a sharp decline in the popularity of the already-present one. This same effect can also be found in the more complex

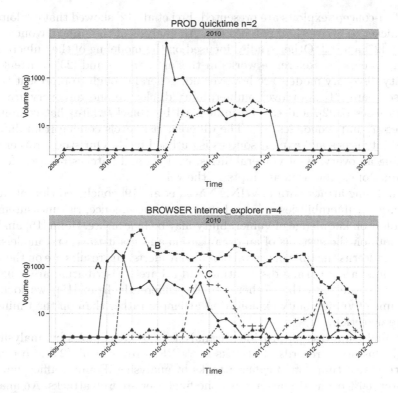

Fig. 5. Trends in attacks for vulnerabilities disclosed in 2010 for QuickTime and Internet Explorer

scenario of Internet Explorer: in 2010 we have three main exploits (a fourth is collapsed several orders of magnitude below the others). Let's call them A (full line, dots), B (dashed lined, squares) and C (short dashes, crosses). We note that:

- when A falls, B rises
- after a sharp decline in B, C rises
- when B goes up again C falls and A disappears
- when B finally dies, first C rises and then A (since then dead) rises up again

We therefore find supporting evidence for Hyp 2. We find Hyp 3 to be supported as well as in both cases one exploit dominates all the others at least by one order of magnitude. We keep a more precise and formal characterization of this model for future research.

8   Related Work

Shahzad et al. [28] have recently presented a general overview of software vulnerabilities. Many descriptive trends in timings of vulnerability patching and release of proof-of-concept exploits are presented. Frei et al. [12] showed that exploits are often quicker to arrive than patches are. An analysis of the same flavour is provided by [27] and [7]. Other studies focused on the modeling of the vulnerability discovery processes. Reference works in this area are [1] and [23]. Current vulnerability discovery models are however not general enough to represent trends for all software [21]. Moreover, vulnerability disclosure and discovery are complex processes [7,22], and can be influenced by {black/white}-hat community activities [7] and economics [17]. The different risk levels coming from different vulnerability types and exploit sources is outlined in [4]. Our study, rather than presenting an overview of vulnerabilities, exploits and patches releases, focuses on volumes of exploitation attempts in the wild.

By analysing attack data in WINE Nayak et al. [19] concluded that attackers focus on few vulnerabilities only and that, as a consequence, risk measurements based solely on knowledge of vulnerability may be inaccurate. Holm [15] analyses attack data on the systems of an organisation and fits it to several models. His analysis concerns the time of arrival of malware alerts. His results are on the same lines as ours: a log-normal distribution and a Pareto distribution are usually a better fit to the data than other models. Differently from [15], we focus on the volume of vulnerability exploitation attempts rather than on the timings of malware detection.

Bilge and Dumitras [6] provide an analysis of 0-day exploits by analysing in hindsight historical records of attacks in WINE. Provos et al. [24] also provide a quantitative estimation of cyber-attacks by analysing iFrame traffic; they find that about 60% of the threats against the final user are web attacks. An analysis of the mechanisms responsible for the generation of these attacks can be found in [14], that uncovers the Exploit-as-a-Service architecture for cyberattacks used by cybercriminals. Following this line of research, an estimation of the fraction of attacks generated by cybercrime market activities is given in [5]. An in-depth, empirical analysis of the tools used by cybercriminals to deliver their attacks is given in [16] and [2]. Rather than focusing on the general volume of attacks affecting the final user, in this work we evaluate how vulnerability exploitation is distributed in the wild.

9   Conclusions

In this paper we analysed the frequency with which vulnerabilities are exploited in the wild. Our findings clearly show that the distribution of attacks follows a heavily tail distribution, showing that a small fraction of vulnerabilities is responsible for the great majority of attacks against a software.

We hypothesise that this distribution may follow a Power Law, but this hypothesis is only inconclusively supported by our evidence: an alternative, equally good explanation to the data may be provided by a log-normal distribution. The statistical power needed to accept one or the other hypothesis is reduced by the relatively low number of vulnerabilities present in our dataset which nonetheless represents, at the best of our knowledge, the most comprehensive collection of attacks in the wild publicly available at the moment of writing.

To further explain our results we present the *Law of the Work-Averse Attacker*, according to which attackers only select one vulnerability to exploit at a time, per software. This results in a distribution of attacks whereby only one vulnerability out of many represent a relevant risk for the user. This model is qualitatively supported by the evidence we find by analysing two case scenarios for Apple Quicktime and Microsoft Internet Explorer. We leave a more formal analysis of this model to future work.

References

1. Alhazmi, O., Malaiya, Y.: Modeling the vulnerability discovery process. In: Proceedings of the 16th IEEE International Symposium on Software Reliability Engineering (ISSRE 2005), pp. 129–138 (2005)
2. Allodi, L., Kotov, V., Massacci, F.: Malwarelab: Experimentation with cybercrime attack tools. In: Proceedings of the 2013 6th Workshop on Cybersecurity Security and Test (2013)
3. Allodi, L., Massacci, F.: A preliminary analysis of vulnerability scores for attacks in wild. In: Proceedings of the 2012 ACM CCS Workshop on Building Analysis Datasets and Gathering Experience Returns for Security (2012)
4. Allodi, L., Massacci, F.: Comparing vulnerability severity and exploits using case-control studies. ACM Transaction on Information and System Security (TIS-SEC) 17(1) (August 2014)
5. Allodi, L., Woohyun, S., Massacci, F.: Quantitative assessment of risk reduction with cybercrime black market monitoring. In: Proceedings of the 2013 IEEE S&P International Workshop on Cyber Crime (2013)
6. Bilge, L., Dumitras, T.: Before we knew it: an empirical study of zero-day attacks in the real world. In: Proceedings of the 19th ACM Conference on Computer and Communications Security (CCS 2012), pp. 833–844. ACM (2012)
7. Clark, S., Frei, S., Blaze, M., Smith, J.: Familiarity breeds contempt: the honeymoon effect and the role of legacy code in zero-day vulnerabilities. In: Proceedings of the 26th Annual Computer Security Applications Conference, pp. 251–260 (2010), http://doi.acm.org/10.1145/1920261.1920299
8. Clauset, A., Shalizi, C.R., Newman, M.E.: Power-law distributions in empirical data. SIAM Review 51(4), 661–703 (2009)
9. Clauset, A., Young, M., Gleditsch, K.S.: On the frequency of severe terrorist events. Journal of Conflict Resolution 51(1), 58–87 (2007), http://jcr.sagepub.com/content/51/1/58.abstract
10. Council, P.: Pci dss requirements and security assessment procedures, version 2.0 (2010), https://www.pcisecuritystandards.org/documents/pci_dss_v2.pdf
11. Efron, B., Tibshirani, R.J.: An introduction to the bootstrap, vol. 57. CRC Press (1994)
12. Frei, S., May, M., Fiedler, U., Plattner, B.: Large-scale vulnerability analysis. In: Proceedings of the 2006 SIGCOMM Workshop on Large-Scale Attack Defense, pp. 131–138. ACM (2006)
13. Gillespie, C.S.: Fitting heavy tailed distributions: the poweRlaw package, package version 0.20.2 (2013)

14. Grier, C., Ballard, L., Caballero, J., Chachra, N., Dietrich, C.J., Levchenko, K., Mavrommatis, P., McCoy, D., Nappa, A., Pitsillidis, A., Provos, N., Rafique, M.Z., Rajab, M.A., Rossow, C., Thomas, K., Paxson, V., Savage, S., Voelker, G.M.: Manufacturing compromise: the emergence of exploit-as-a-service. In: Proceedings of the 19th ACM Conference on Computer and Communications Security (CCS 2012), pp. 821–832. ACM (2012)
15. Holm, H.: A large-scale study of the time required to compromise a computer system. IEEE Transactions on Dependable and Secure Computing 11(1), 2–15 (2014)
16. Kotov, V., Massacci, F.: Anatomy of exploit kits. In: Jürjens, J., Livshits, B., Scandariato, R. (eds.) ESSoS 2013. LNCS, vol. 7781, pp. 181–196. Springer, Heidelberg (2013)
17. Miller, C.: The legitimate vulnerability market: Inside the secretive world of 0-day exploit sales. In: Proceedings of the 6th Workshop on Economics and Information Security (2007)
18. Mitzenmacher, M.: A brief history of generative models for power law and lognormal distributions. Internet Mathematics 1(2), 226–251 (2004)
19. Nayak, K., Marino, D., Efstathopoulos, P., Dumitraş, T.: Some vulnerabilities are different than others. In: Stavrou, A., Bos, H., Portokalidis, G. (eds.) RAID 2014. LNCS, vol. 8688, pp. 426–446. Springer, Heidelberg (2014)
20. Newman, M.E.: Power laws, pareto distributions and zipf's law. Contemporary Physics 46(5), 323–351 (2005)
21. Nguyen, V.H., Massacci, F.: An independent validation of vulnerability discovery models. In: Proceeding of the 7th ACM Symposium on Information, Computer and Communications Security, ASIACCS 2012 (2012)
22. Ozment, A.: The likelihood of vulnerability rediscovery and the social utility of vulnerability hunting. In: Proceedings of the 4th Workshop on Economics and Information Security (2005)
23. Ozment, A.: Improving vulnerability discovery models: Problems with definitions and assumptions. In: Proceedings of the 3rd Workshop on Quality of Protection (2007)
24. Provos, N., Mavrommatis, P., Rajab, M.A., Monrose, F.: All your iframes point to us. In: Proceedings of the 17th USENIX Security Symposium, pp. 1–15 (2008)
25. Quinn, S.D., Scarfone, K.A., Barrett, M., Johnson, C.S.: Sp 800-117. guide to adopting and using the security content automation protocol (scap) version 1.0. Tech. rep., National Institute of Standards & Technology (2010)
26. R Core Team: R: A Language and Environment for Statistical Computing. R Foundation for Statistical Computing, Vienna, Austria (2013), http://www.R-project.org
27. Ransbotham, S.: An empirical analysis of exploitation attempts based on vulnerabilities in open source software. In: Proceedings of the 9th Workshop on Economics and Information Security (2010)
28. Shahzad, M., Shafiq, M.Z., Liu, A.X.: A large scale exploratory analysis of software vulnerability life cycles. In: Proceedings of the 34th International Conference on Software Engineering, pp. 771–781. IEEE Press (2012)
29. Vuong, Q.H.: Likelihood ratio tests for model selection and non-nested hypotheses. Econometrica: Journal of the Econometric Society, 307–333 (1989)
30. Wash, R.: Folk models of home computer security. In: Proceedings of the Sixth Symposium on Usable Privacy and Security (2010)

Idea: Benchmarking Indistinguishability Obfuscation – A Candidate Implementation

Sebastian Banescu, Martín Ochoa, Nils Kunze, and Alexander Pretschner

Technische Universität München, Germany
{banescu,ochoa,nils.kunze,pretschn}@cs.tum.edu

Abstract. We present the results of preliminary experiments implementing the Candidate Indistinguishability Obfuscation algorithm recently proposed by Garg et al. [1]. We show how different parameters of the input circuits impact the performance and the size of the obfuscated programs. On the negative side, our benchmarks show that for the time being the algorithm is far away from being practical. On the positive side, there is still much room for improvement in our implementation. We discuss bottlenecks encountered and optimization possibilities. In order to foster further improvements by the community, we make our implementation public.

1 Introduction

Obfuscation of software, intended as a transformation of a program such that it is difficult for adversaries to understand details of its logic or internal variables, is of an increasing practical relevance [2]. Typically obfuscation is associated with 'security by obscurity', because of a lack of formal guarantees on the security of commonly used obfuscation operators [3]. On the theoretical side, it has been shown by Barak et al. [4] that it is impossible to construct an obfuscator such that from the obfuscated version of a program implementing a function f, an adversary can only learn the inputs and outputs to f exclusively.

Recently, Garg et al. [1] proposed a promising approach that offers formal security guarantees for *indistinguishability obfuscation*, a particular obfuscation notion that guarantees that the obfuscations of two programs implementing the same functionality are indistinguishable. As a basis for their proof, the authors show that a successful attack to their construction is also an solution to the multilinear jigsaw problem, which is believed to be computationally hard. The authors conjecture that this construction provides the expected security for the obfuscation of most programs.

Although the proposers of indistinguishability obfuscation acknowledge that their construction is not practical as of today [5], concrete details have so far not been published. The motivation of our work is thus to better understand how far is the candidate construction from real applications. To do so, we prototypically implemented the algorithm described in [1] and benchmarked its space and time performance depending on various parameters.

F. Piessens et al. (Eds.): ESSoS 2015, LNCS 8978, pp. 149–156, 2015.

Our contributions can be summarized as follows: a) to our knowledge, we provide the first open source implementation of the candidate indistinguishability obfuscation candidate [1], so that the community can gradually improve on it, b) we provide reproduceable performance benchmarks, which give an upper bound on the necessary time and space for running/storing obfuscated programs and c) we discuss potential areas for improvement based on our experiments.

The paper is organized as follows: In Section 2 we give an overview of the candidate construction. We then present an overview of our implementation in Section 3 and our benchmarking results in Section 4. We conclude by summarizing our results and giving an overview of ongoing and future work in Section 5.

2   Preliminaries

This section presents the candidate indistinguishability obfuscation construction developed by Garg *et al.* [1] applied to *boolean circuits* in $\mathbf{NC}^1$ [6], preceded by the concepts needed to understand this construction.

A boolean circuit is a directed acyclic graph, where nodes are represented by conjunction, disjunction and/or negation gates with maximum 2 inputs (fan-in-2), which process only boolean values. The *size* of a circuit is equal to the total number of gates in that circuit. The *depth* of a circuit is the length of the longest path from input to output gate, in the circuit.

A uniform probabilistic polynomial-time Turing (PPT) machine $i\mathcal{O}$ is called an *indistinguishability obfuscator* for a circuit class $\{\mathcal{C}_\lambda\}$ if: (1) it preserves the input-output behavior of the unobfuscated circuit and (2) given two circuits $C_1, C_2 \in \mathcal{C}_\lambda$ and their obfuscated counterparts $i\mathcal{O}(\lambda, C_1), i\mathcal{O}(\lambda, C_2)$, a PPT adversary will not be able to distinguish which obfuscated circuit originates from which original circuit with significant probability (the advantage of the adversary is bounded by a negligible function of the security parameter λ).

Even though an $i\mathcal{O}$ applies to boolean circuits, internally it transforms all circuits into *linear branching programs* on which it operates. This transformation is made possible by Barrington's theorem [7], which states that any fan-in-2, depth-d boolean circuit can be transformed into an oblivious linear branching program of length at most 4^d, that computes the same function as the circuit.

Definition 1. *(Oblivious Linear Branching Program [1]) Let $A_0, A_1 \in \{0,1\}^{5 \times 5}$ be two distinct arbitrarily chosen permutation matrices. An (A_0, A_1) oblivious branching program of length n for circuits with ℓ-bit inputs is represented by a sequence of instructions $BP = ((inp(i), A_{i,0}, A_{i,1}))_{i=1}^n$, where $A_{i,b} \in \{0,1\}^{5 \times 5}$, and $inp: \{1, n\} \to \{1, \ell\}$ is a mapping from branching program instruction index to circuit input bit index. The function computed by the branching program is*

$$f_{BP,A_0,A_1}(x) = \begin{cases} 0 & \text{if } \Pi_{i=1}^n A_{i,x_{inp(i)}} = A_0 \\ 1 & \text{if } \Pi_{i=1}^n A_{i,x_{inp(i)}} = A_1 \\ \text{undef } \text{otherwise} \end{cases}$$

The family of circuits $\mathcal{C}_\lambda$ is characterized by ℓ inputs, λ gates, $O(log\lambda)$ depth and one output. $\mathcal{C}_\lambda$ has a corresponding polynomial-sized universal circuit, which is a function $U_\lambda : \{0,1\}^{f(\lambda)} \times \{0,1\}^\ell \to \{0,1\}$, where $f(\lambda)$ is some function of λ. U_λ can encode all circuits in $\mathcal{C}_\lambda$, i.e. $\forall C \in \mathcal{C}_\lambda, \forall z \in \{0,1\}^\ell, \exists C_b \in \{0,1\}^{f(\lambda)}$: $U_\lambda(C_b, z) = C(z)$. It is important to note that the input of U_λ is a $f(\lambda) + \ell$ bit string and that by fixing any $f(\lambda)$ bits, one obtains a circuit in $\mathcal{C}_\lambda$.

Universal circuits are part of the candidate $i\mathcal{O}$ construction, because they enable running Kilian's protocol [8], which allows two parties (V and E), to evaluate any $\mathbf{NC}^1$ circuit (e.g. U_λ) on their joint input $\mathcal{X} = (x|y)$, without disclosing their inputs to each other, where x, y are the inputs of V, respectively E. This is achieved by transforming the circuit into a branching program $BP = ((inp(i), A_{i,0}, A_{i,1}))_{i=1}^n$ by applying Barrington's theorem [7]. Subsequently V chooses n random invertible matrices $\{R_i\}_{i=1}^n$ over $\mathcal{Z}_p$, computes their inverses and creates a new *randomized branching program* $RBP = ((inp(i), \tilde{A}_{i,0}, \tilde{A}_{i,1}))_{i=1}^n$, where $\tilde{A}_{i,b} = R_{i-1} A_{i,b} R_i^{-1}$ for all $i \in \{1, n\}, b \in \{0, 1\}$ and $R_0 = R_n$. It can be shown that RBP and BP compute the same function. Subsequently, V sends E only the matrices corresponding to her part of the input $\{\tilde{A}_{i,b} : i \in \{1, n\}, inp(i) < |x|\}$ and E only gets the matrices corresponding to one specific input via oblivious transfer. E can now compute the result of RBP without finding out V's input. Kilian's protocol is related to the notion of program obfuscation, if we think of V as a software vendor who wants to hide (obfuscate) a program that is going to be distributed to end-users (E). However, Kilian's protocol [8] is modified in [1], by sending all matrices corresponding to any input of E, which allows E to run the RBP with more that one input. This modified version is vulnerable to *partial evaluation attacks*, *mixed input attacks* and also non-multilinear attacks, which extract information about the secret input of V.

To prevent partial evaluation attacks Garg *et al.* [1] transform the 5×5 matrices of BP into higher order matrices, having dimension $2m + 5$, where $m = 2n + 5$ and n is the length of BP. Subsequently, they add 2 *bookend* vectors of size $2m + 5$ in order to neutralize the multiplication with the random entries in the higher order matrices. To prevent mixed input attacks a multiplicative bundling technique is used, which leads to an *encoded* output of BP. To decode the output of the BP an additional branching program of equal length with BP, that computes the constant 1 function is generated and the same multiplicative bundling technique is applied to it. Subtracting the results of the two branching programs executed on the same inputs, will decode the output of BP. To prevent non-multilinear attacks, the candidate construction of Garg *et al.* [1] employs the *multilinear jigsaw puzzle* (MJP).

An overview of MJP is illustrated in Figure 1 and consists of two entities, i.e. the *Jigsaw Generator* (JGen) and the *Jigsaw Verifier* (JVer). The JGen is part of the *circuit obfuscator*. It takes as input a security parameter (λ), a universal circuit (U_λ) and the number of input bits (ℓ) of any circuit simulated by U_λ. JGen first applies Barrington's theorem [7] to transform U_λ into a universal branching program UBP of length n. Subsequently, the *Instance Generator* takes λ and the multilinearity parameter ($k = n+2$) as inputs and outputs a prime number p and a set of public system parameters (including a large random prime q and a small random polynomial $g \in \mathcal{Z}[X]/(X^m + 1)$). Afterwards, UBP is transformed into

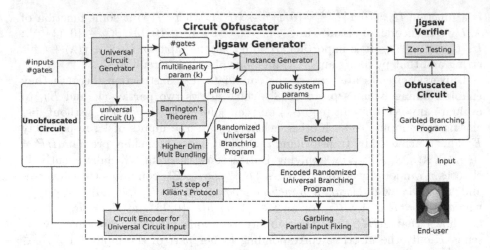

Fig. 1. Overview of the candidate construction for indistinguishability obfuscation

a randomized branching program by: (1) transforming the branching program matrices into higher order matrices, (2) applying multiplicative bundling and (3) the first step of Kilian's protocol. The output of JGen is a set of public system parameters and the randomized universal branching program ($\widehat{\mathcal{RND}}(UBP_\lambda)$) with all matrices encoded by the *Encoder* component.

The output of JGen can be used to obfuscate a circuit $C \in \mathcal{C}_\lambda$ by fixing a part of the inputs (garbling) of $\widehat{\mathcal{RND}}(UBP_\lambda)$ such that it encodes C for all $z \in \{0,1\}^\ell$. Garbling is done by discarding the matrices of $\widehat{\mathcal{RND}}(UBP_\lambda)$ which correspond to values not chosen for the fixed input bits. The result of this step is $i\mathcal{O}(\lambda, C)$, the candidate of Garg *et al.* [1]. It is sent to an untrusted party which evaluates it by fixing the rest of its inputs and providing it as input to the JVer. The JVer outputs 1 if the evaluation of $i\mathcal{O}(\lambda, C)$ is successful and 0, otherwise.

3   Implementation

Our proof-of-concept implementation was done in Python, leveraging the SAGE computer algebra system and can be downloaded from the Internet[1]. It consists of the following modules, corresponding to the light blue rectangles from Figure 1: (1) building blocks for universal circuit creation, (2) Barrington's theorem for transforming boolean circuits to branching programs, (3) transformation from branching program matrices into higher order matrices and applying multiplicative bundling (4) 1st step of Kilian's protocol for creating randomized branching programs from branching programs, (5) instance generator for MJP, (6) encoder for MJP, (7) circuit encoder into input for universal circuit, (8) partial input fixer for random branching programs, and (9) zero testing of jigsaw verifier.

---

[1] https://github.com/tum-i22/indistinguishability-obfuscation

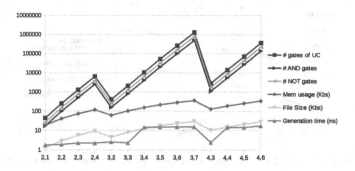

Fig. 2. Generation of UCs (X-axis: no. inputs (ℓ), no. gates of input circuit (λ))

Technical Challenges Faced. Although commonly used in the literature, we could not find a readily available implementation of Universal Circuits (UC) that was easily adaptable to our setting. Therefore we decided to implement our own UC component, following the less performant algorithm of [9]. For the sake of performance, this component can be improved by following for instance the more performant (but more complex) algorithm suggested in [9] or [10].

Challenges Interpreting [1]. We also faced some challenges while interpreting the candidate construction description, in particular their suggested encoding function. For instance it was difficult to come up with concrete values for some parameters, since the relation between them is given using the big O notation. On the other hand, the Encoder function requires to reduce an element $a \in \mathcal{Z}_p$ modulo a polynomial g of degree ≥ 1. We could not think of a better canonical representative for this reduction than a itself, which makes us believe that either the modulo reduction is redundant or the authors had another canonical representative in mind (a polynomial) which is unclear how to compute.

Summary of Current Status. Currently, our implementation can perform most steps of the candidate construction, with the exception of the zero test. We believe this is a result of an incorrect choice of the canonical representative of a modulo g or/and of the concrete parameters as discussed above. We have raised these issues in popular mathematics and cryptography forums and contacted the authors for clarification with no success at the moment of elaborating this document. However, note from Figure 1 that the improper functioning of the zero test does not affect the results of benchmarking the *Circuit Obfuscator* presented in the next section, because the it is part of the *Jigsaw Verifier*.

4   Benchmarking

We executed our experiments on a virtual machine (VM) with 4 cores and 64 GB of memory. The first experiment aims to investigate the resources required

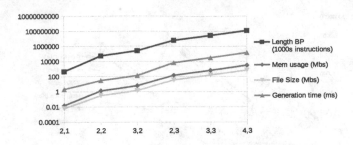

Fig. 3. Generation of BPs (X-axis: no. inputs (ℓ), no. gates of input circuit (λ))

to obfuscate a circuit consisting only of AND gates as a function of its number of inputs and gates. As illustrated in Figure 1 the first step of obfuscation consists of generating the UC, corresponding to the first step of our experiment. The number of circuit inputs were varied between 2 and 4, while the number of gates between 1 and 10. The recorded outputs are shown in Figure 2 and consist of the: number of gates, memory usage, output file and generation time needed for the UC. Observe that increasing the number of inputs causes a linear increase in each measured output of the experiment, while increasing the number of gates causes an exponential increase. The memory usage is around one order of magnitude higher than the file size due to the compression algorithm we use to store UCs.

The second step of our experiment consisted of transforming the previously generated UCs into branching programs (BPs) using our implementation of Barrington's theorem [7]. However, it was infeasible to transform all the previously generated UCs because of the fast polynomial increase in memory usage and file size, illustrated in Figure 3. We estimated the size of generating a BP for a UC which encodes a circuit by applying following recursive formula (corresponding to our implementation), to the output gate of a UC:

$$l(\text{gate}) = \begin{cases} 1 & \text{if type(gate)} = \mathsf{Input} \\ l(\text{gate.input}) & \text{if type(gate)} = \mathsf{NOT} \\ 2l(\text{gate.input1}) + 2l(\text{gate.input2}) & \text{if type(gate)} = \mathsf{AND} \end{cases}$$

The estimated memory usage of a universal BP which encodes 4 inputs and 6 gates, corresponding to the largest UC we show in Figure 2, is over 4.47 Peta Bytes, which is infeasible to generate on our VM.

The third step of our experiment was to transform the BPs generated previously into randomized branching programs (RBPs) by transforming the BP matrices into higher order matrices, applying multiplicative bundling and the first step of Kilian's protocol [8]. The results of this experiment are shown in Figure 4. Additionally to the number of inputs and gates, in this experiment we also have the matrix dimension increase (m) and the choice of the prime (p) corresponding to $\mathcal{Z}_p$ in which Kilian's protocol operates. The choice of m influences both the generation time and the file size polynomially. Observe that the memory usage remains constant for different values of m. This is due to compatibility issues between SAGE and our memory profiler. However, we observer

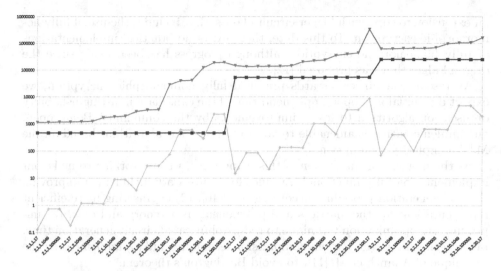

Fig. 4. Generation of RBPs (X-axis: no. inputs (ℓ), no. gates of input circuit (λ), matrix dimension (m), prime number (p)). Legend is the same as Figure 3.

that the actual memory usage is still one order of magnitude higher than the file size. p influences the generation time linearly, however, the memory usage and file size are affected only if the data type width of p grows. Note that, the memory usage is not shown in Figure 4 since it could not be measured reliably due to technical limitations of our memory profiler. We estimate that the memory usage is approximately one order of magnitude higher than the file size.

5   Conclusions and Future Work

In this paper we have presented a non-trivial upper bound on the size and performance of the obfuscated versions of small circuits. To give an idea about the practicality of this construction, consider a 2-bit multiplication circuit. It requires 4 inputs and between 1 and 8 AND gates for each of its 4 output bits. An obfuscation would be generated in about 10^{27} years on a 2,6 GHz CPU and would require 20 Zetta Bytes of memory for $m = 1$ and $p = 1049$. Executing this circuit on the same CPU would take 1.3×10^8 years. This clearly indicates that for the time being the candidate construction is highly unpractical.

However, this upper bound can still be tightened (perhaps even dramatically) by improving upon our preliminary implementation. In particular, there exist better algorithms for the generation of universal circuits, which directly affect the size of the obfuscation [9, 10]. There is an inherent limitation for this improvement due to the fact that the output of gates in UCs are reused by other gates, which causes duplication of matrices in BPs when using Barrington's theorem [7]. Therefore, one improvement is to avoid using Barrington's theorem as suggested by Ananth et al. [11]. On the other hand, we have only implemented the construction for $\mathbf{NC}^1$ circuits: the candidate construction includes an extension to cope with bigger circuit classes, that includes the use of fully homomorphic encryption. To this date, there exists no practical implementations of fully homomorphic encryption, although progress has been made since the original algorithm was proposed [12].

As research advances towards practical fully homomorphic encryption, we expect our initial and open implementation of the candidate indistinguishability obfuscation algorithm to foster improvements by the community. Being open, our implementation is amenable to adaptations to new algorithms based on the MJP complexity assumption.

At the moment of submission of this manuscript, we are working to make our implementation fully functional. Avenues for future work include: (1) improving the UC generation procedure according to [9,10], (2) engineering more efficient representations for the matrices and polynomials in memory and disk, (3) improving our optimization technique to reduce obfuscated circuit generation time, (4) experimenting with various compression techniques and (5) implementing the technique of Ananth et al. [11], to avoid Barrington's theorem.

References

1. Garg, S., Gentry, C., Halevi, S., Raykova, M.: A Sahai, and B. Waters. Candidate indistinguishability obfuscation and functional encryption for all circuits. In: Proc. of the 54th Annual Symp. on Foundations of Computer Science, pp. 40–49 (2013)
2. Zhou, W., Wang, Z., Zhou, Y., Jiang, X.: Divilar: Diversifying intermediate language for anti-repackaging on android platform. In: Proc. of the 4th ACM Conf. on Data and Application Security and Privacy, pp. 199–210. ACM (2014)
3. Collberg, C., Thomborson, C., Low, D.: A taxonomy of obfuscating transformations. Technical report, Department of Computer Science, The University of Auckland, New Zealand (1997)
4. Barak, B., Goldreich, O., Impagliazzo, R., Rudich, S., Sahai, A., Vadhan, S.P., Yang, K.: On the (Im)possibility of obfuscating programs. In: Kilian, J. (ed.) CRYPTO 2001. LNCS, vol. 2139, pp. 1–18. Springer, Heidelberg (2001)
5. Edwards, C.: Researchers probe security through obscurity. Communications of the ACM 57(8), 11–13 (2014)
6. Arora, S., Barak, B.: Computational complexity: a modern approach. Cambridge University Press (2009)
7. Barrington, D.A.: Bounded-width polynomial-size branching programs recognize exactly those languages in nc1. In: Proc. of the 18th Annual ACM Symp. on Theory of Computing, STOC 1986, pp. 1–5. ACM, New York (1986)
8. Kilian, J.: Founding crytpography on oblivious transfer. In: Proc. of the 20th Annual ACM Symp. on Theory of Computing, pp. 20–31. ACM (1988)
9. Schneider, T.: Practical secure function evaluation. Master's thesis, Friedrich-Alexander-Universität Erlangen-Nürnberg (2008)
10. Valiant, L.G.: Universal circuits (preliminary report). In: Proc. of the 8th Annual ACM Symp. on Theory of Computing, pp. 196–203. ACM (1976)
11. Ananth, P., Gupta, D., Ishai, Y., Sahai, A.: Optimizing obfuscation: Avoiding barrington's theorem. IACR Cryptology ePrint Archive 2014, 222 (2014)
12. Naehrig, M., Lauter, K., Vaikuntanathan, V.: Can homomorphic encryption be practical? In: Proc. of the 3rd ACM Workshop on Cloud Computing Security, pp. 113–124. ACM (2011)

A Security Ontology for Security Requirements Elicitation

Amina Souag[1], Camille Salinesi[1], Raúl Mazo[1], and Isabelle Comyn-Wattiau[2]

[1] CRI -Paris 1 Sorbonne University
Paris, France
{amina.souag,camille.salinesi,raul.mazo}@univ-paris1.fr
[2] CEDRIC-CNAM & ESSEC Business School
Paris, France
isabelle.wattiau@cnam.fr

Abstract. Security is an important issue that needs to be taken into account at all stages of information system development, including early requirements elicitation. Early analysis of security makes it possible to predict threats and their impacts and define adequate security requirements before the system is in place. Security requirements are difficult to elicit, analyze, and manage. The fact that analysts' knowledge about security is often tacit makes the task of security requirements elicitation even harder. Ontologies are known for being a good way to formalize knowledge. Ontologies, in particular, have been proved useful to support reusability. Requirements engineering based on predefined ontologies can make the job of requirement engineering much easier and faster. However, this very much depends on the quality of the ontology that is used. Some security ontologies for security requirements have been proposed in the literature. None of them stands out as complete. This paper presents a core and generic security ontology for security requirements engineering. Its core and generic status is attained thanks to its coverage of wide and high-level security concepts and relationships. We implemented the ontology and developed an interactive environment to facilitate the use of the ontology during the security requirements engineering process. The proposed security ontology was evaluated by checking its validity and completeness compared to other ontologies. Moreover, a controlled experiment with end-users was performed to evaluate its usability.

Keywords: Security, ontology, concepts, security requirements, elicitation.

1   Introduction

Security has moved from being considered by Information Systems (IS) designers as a technical topic to becoming a critical issue in our society [1]. With the growing digitization of activities, IS are getting more and more complex. They must comply with new usages, varied needs, and are permanently exposed to new vulnerabilities. There is no single week without an announcement indicating that the IS of some

F. Piessens et al. (Eds.): ESSoS 2015, LNCS 8978, pp. 157–177, 2015.

private or public organization was attacked. The cost of cybercrime in 2012 reached $110B in the world [2]. It has been reported recently that attacks to sensitive data increased by 62% in 2013 with 253 incidents observed and 552 million identities stolen [44]. A major obstacle that faces analysts, and requirements engineers, is the fact that knowledge about security is most often tacit, imprecisely defined and non-formalized. Among the challenges for security projects is the difficulty of expressing security requirements and producing exhaustive specifications. A requirement prescribes a condition judged necessary for the system [3]. Security Requirements Engineering (SRE) methods derive security requirements using specific concepts, borrowed from security engineering paradigms [4]. It is well known that ontologies are useful for representing and inter-relating many types of knowledge of a same domain. Thus, the research community of information system security [5] urged the necessity of having a good security ontology to harmonize the vaguely defined terminology, leading to communication troubles between stakeholders. The benefits of such a security ontology would be manifold: it would help requirements engineers reporting incidents more effectively, reusing security requirements of the same domain and discussing issues together, for instance [6]. Several research studies have addressed the issue of knowledge for the field of security [7][8]. The research presented in this paper is part of a larger ongoing research project that aims at proposing a method that exploits ontologies for security requirements engineering [9]. In [9], a small security ontology was first used for the elicitation and analysis of security requirements. Being "small", the ontology used affected the resulting requirements and the whole security requirements analysis process. In a previous research, several security ontologies were compared and classified [7]. The paper concluded that ontologies are good sources for security requirements engineering. However the quality of the resulting security requirements depends greatly on the ontologies used during the elicitation and analysis process. To cope with the aforementioned issues, this paper proposes a core security ontology that considers the descriptions of the most important concepts related to security requirements and the relationships among them. *"Core"* refers to the union of knowledge (high-level concepts, relationships, attributes) present in other security ontologies proposed in the literature. As Massacci et al. claims, "Although there have been several proposals for modeling security features, what is still missing are models that focus on high-level security concerns without forcing designers to immediately get down to security mechanisms"[15]. Meta-models can be useful since they provide an abstract syntax of security concepts. However, we believe that ontologies can be a better option since they allow representing, accessing, using and inferring about that knowledge in order to develop methods, techniques, and tools for security requirements analysis. According to [16], a good security ontology should inter alia include static knowledge (concepts, relationships and attributes), and dynamic knowledge (axioms). It must be reusable (commented in natural language, and formalized in a standard language). The main objective of this paper is to address the following research questions: *What are the concepts and relations that need to be present in a core security ontology? And how to make this ontology easy for requirements engineers to use?* This ontology should make it possible to: (a) Create a generic platform of different security concepts (threats, risks, requirements, etc.). (b) Create a source of reusable knowledge for the elicitation of security requirements in various projects.

The rest of the paper is organized as follows: Section 2 presents the construction of the ontology, its concepts and relationships. Section 3 reports the evaluation of the proposed ontology. Related works are presented in Section 4 through a literature review. Finally, Section 5 concludes the paper and describes future work directions.

2   A Core Security Ontology for Security Requirements Engineering

This section presents the main contribution of this paper, a core security ontology to be used particularly for security requirements elicitation process. The method for constructing the security ontology is adapted from ontology construction methods proposed by Fernandez [25], mixed with key principles of the ones proposed by Jones et al. [26]. The construction process contains six main steps: objective, scope, knowledge acquisition, conceptualization, implementation, and validation. The objective behind the ontology construction must be defined in the beginning, including its intended uses, scenarios of use, end-users, etc. The scope stipulates the field covered by the ontology. The knowledge acquisition step aims at gathering from different sources the knowledge needed for the ontology construction. In the step of conceptualization, the knowledge is structured in a conceptual model that contains concepts and relationships between them. Ontology implementation requires the use of a software environment such as Protégé[1]; this includes codifying the ontology in a formal language (RDF or OWL/XML). Finally, the validation step guarantees that the resulting ontology corresponds to what it is supposed to represent. The details about how the first five steps were applied to construct our ontology are presented in the following sub-sections and the last step is detailed in Section 3.

2.1   Objective

The main objective of the target ontology is to provide a generic platform containing knowledge about the core concepts related to security (threats, vulnerabilities, countermeasures, requirements, etc.). This ontology will be a support for the elicitation of security requirements and the development of SRE methods and tools. The ontology will be a meta-view for the different security ontologies in the literature. It should harmonize the security terminology spread in these ontologies and help requirements engineers communicate with each other.

2.2   Scope of the Ontology

The ontology covers the security domain in its high level aspects (threats and treatments) as well as its organizational ones (security procedures, security management process, assets, and persons). The reader will find details on all security concepts covered by the ontology in section 2.4.below on.

---

[1] http://protege.stanford.edu/

2.3   Knowledge Acquisition

The acquisition of the security knowledge started from standards (e.g. ISO27000). Other knowledge acquisition sources were the different security ontologies that exist in the literature. We analyzed about 20 security ontologies, based on previous literature surveys; the full list of these ontologies can be found in [7] and [8]. These ontologies are of various levels (general, specific, for a particular domain). Relevant concepts and relationships were extracted through a systematic and syntactic analysis of the security ontologies (their concepts and relations). Table 8 in the appendix presents part of them (13 ontologies). For the sake of space, we cannot provide the reader with the description of all the ontologies used as a source of knowledge for the ontology. Brief descriptions of some of them are presented in the following:

- The ISSRM model [27] (top left in Fig. 1.) was defined after a survey of the risk management standards, security related standards, and security management methods. The three groups of concepts proposed in the ISSRM model (asset related concepts, risk related concepts, and risk treatment related concepts) were used to define the three dimensions of the ontology (organization, risk, treatment).
- Fenz et al. [24] have proposed an ontology to model the information security domain. We reused some concepts and relationships of that ontology, in particular the ones related to the infrastructure of organizations (assets, organization), the relationships between threats and assets, and between threats and vulnerabilities. We also reused some standard controls used in Fenz's ontology to define our security requirements.
- Lashras et al.'s security requirements ontology [12] was useful to define the security requirements in our ontology.

Fig. 1 schematizes the knowledge acquisition step and part of the conceptualization phase, starting with the knowledge sources (the different ontologies), the concept

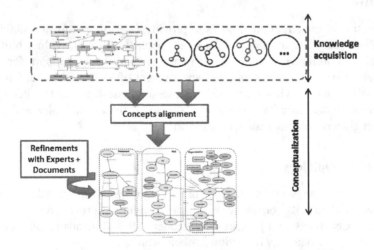

Fig. 1. Knowledge acquisition and conceptualization phases

alignment, and the conceptualization with the help of experts and documents. The concepts of the resulting ontology were derived from the alignments of the different security ontologies in the knowledge acquisition step. The knowledge and the conceptualization steps were performed manually relying essentially on tables to align the concepts and relations of the different source ontologies.

2.4   Conceptualization

Based on the outcomes of the knowledge acquisition step, concepts were organized and structured in a glossary. Various relationships among these concepts were considered, and then were put together in a conceptual model of the ontology (Fig. 4 in the appendix), easy to understand, independently of any implementation language. The names of the concepts and the relationships of the security ontology proposed in this paper were chosen according to the number of occurrences of names in the source ontologies (Table 8 in the appendix). If a concept has different names in the ontologies (e.g. impact or consequence, attack method or deliberate attack, or SessionIP attack); the most generic or easiest to understand name was chosen (here, impact, attack method). Some security experts (5 experts) were consulted to validate the choices that were made. The validation was informal and took the form of email exchanges, phone and direct discussions. The experts acknowledged most of the concepts and relationships between them. Some refinements in the ontology were performed after discussion with them. For example, the concept of "Attack" was removed, since the experts consider it as an Intentional Threat. Discussions also clarified the difference between the concepts of "Security Goal", "Security Criterion", "Security Requirement" and "Control". These concepts are frequently mixed up in the security requirements elicitation phase and the difference between them is often not easy to capture. The concepts were organized around three main dimensions. The latter are: Risk dimension, Treatment dimension, and Organization dimension. In ontology engineering terms [45]: the Risk, Treatment and Organization dimensions are considered as modules. The Risk dimension represents the "dark" face of security; it gathers concepts related to threats, vulnerabilities, attacks, and threat agents. Treatment dimension is concerned with concepts related to the necessary treatments to overcome risks. The concepts are security goals, requirements, controls, and security policies. Finally, security is a multifaceted problem; it is not only about technical solutions or single assets, but also about the environment where threats appear and arise. That is why the Organization dimension is considered. This dimension relates to concepts such as person, location, assets, and organization that must be analyzed and on which assumptions must be match in a security requirements elicitation process. Some ontologies covered only the dimension treatment [12]. The security ontology proposed by Fenz et al. [24] groups concepts into three sets (security, enterprise and location). The classification into these three dimensions (organization, risk and treatment) helps in organizing the knowledge related to security; it has been inspired by the security meta-model proposed in [27]. The concepts and relationships of the ontology are described in the following sub-section. To visualize the different concepts and relations, the reader may refer to Fig.4.

1)   Concepts of the Security Ontology

The following summarizes the different concepts identified for the ontology with their respective descriptions. These general concepts together with their relations constitute the ontology, which presents an overview of the information security in a context-independent manner. In the following, we describe the concepts dimension by dimension.

a) Organization dimension: This dimension includes the concepts related to the organization, its assets and its environment. The concepts are:
Organization: a structure including human, hardware, and software resources (assets).
Person: Represents human agents. A person may be internal in the organization (e.g., administrator) or external (e.g., customer).
Asset: a valuable resource, which can be a tangible asset (e.g., air-conditioning, fire extinguisher, computers) or an intangible asset. Intangible assets can be, for example, software, data, and industrial manufacturing processes.
Location: Defines the asset's location. Location can be a brick and mortar physical location such as a classroom, data center or office. It can also consist of collaborative research materials on a file share or financial information stored in a database [28].

b) Risk dimension: The concepts of the risk dimension are:
Risk: a combination of a vulnerability and threat causing harm to one or more asset.
Severity: the level of risk, e.g. high, medium or low.
Threat: a violation of a security criterion. The threat may be natural, accidental, or intentional (attack).
Vulnerability: a weakness of an asset or group of assets that can be exploited by one or more threats [29] (e.g., weak password).
Impact: the impact may vary from a simple loss of availability to loss of the entire information system control. Impact can also be of other types such as harm to the image of the company.
Threat agent: the person (or program) who carries out the threat. The name 'threat agent' was chosen to cover both types of threat, either intentional (carried out by an attacker) or unintentional (carried out by any person, not necessarily an attacker).
Attack method: Refers to the different methods used by threat agents to accomplish their attacks, such as sniffing (which lets threat agents capture and analyze traffic transmitted over a network); spoofing (where the threat agent attempts to impersonate someone or something else); and social engineering (tricking people into giving sensitive information or performing actions on behalf of the threat agent).
Attack tool: The tool used to perform the attack, e.g. sniffing tool (e.g., Wireshark[2]), spoofing tool (e.g. Subterfuge[3]), scan port tool (e.g., Nmap[4]) and others.

---

[2] http://www.wireshark.org/
[3] http://code.google.com/p/subterfuge/downloads/list
[4] http://nmap.org/

c) Treatment dimension:

Security goal: a security goal defines what a stakeholder/organization hopes to achieve in the future in terms of security [27], it states the intention to counter threats and satisfy security criteria. Security goals are sometimes considered as security objectives [47].

Security Requirement: a condition defined on the environment that needs to be fulfilled in order to achieve a security goal and mitigate a risk. Depending on what we want to protect and on the target security level, we define our requirements. They can be related to databases, applications, systems, organizations, and external environments. For example, "the system shall ensure that personal data can be accessed only by authorized users" and "the system shall deliver data in a manner that prevents further or second hand use by unauthorized people".

Control: a means or a way to secure assets and enable a security requirement, e.g., alarm or password.

Security criterion: defines security properties such as confidentiality, integrity, availability, and traceability. It can also be considered as a constraint on assets.

Requirements document: The document that states in writing the necessary security requirements to protect the assets. Two main documents generally contain security requirements:

  - Security policy: a security policy expresses the defense strategy or strategic directions of the information security board of an organization.

  - Specification document: it gathers the set of requirements to be satisfied by a material, design, product, or service. The document contains, inter alia, security requirements.

2)   Relationships of the Security Ontology

High-level relationships between those concepts were defined. They were categorized into four kinds: IsA, HasA, SubClassOf and AssociatedTo. The relationships between the concepts of the security ontology can be briefly described as follows: An organization has assets (Has_Asset). An asset may have a location (Has_Location). Tangible and intangible assets are subclasses of the asset concept (SubClassOf). An organization also includes persons that it deals with (Has_Person). The persons can be internal or external (SubClassOf). An asset is threatened by one or many threats (Threatens). These threats exploit vulnerabilities in the assets (Exploits). The threat-agent leads an attack (LeadBy) and uses attack methods (UseMethod) or attack tools (UseTool) to achieve an attack. A threat implies an impact (Implies), for example: "A denial of service attack implies a server downtime". The impact affects one or more assets (Affect). A threat can be natural, intentional, or accidental (SubClassOf). A threat generates a risk (Generate) with a certain level of severity (HasSeverity). Security requirements mitigate a risk (Mitigate) and satisfy (Satisfy) security goals expressed by stakeholders (ExpressedBy). Security requirements fulfill (Fulfills) one or more security criteria. For instance, the requirement "The application shall ensure that each user will be able to execute actions for which he/she has permission at any time/every week" satisfies the security criteria Confidentiality and Availability. Controls enable a security requirement (Enable). For example, the control "password" enables the requirement "The application shall ensure that each user will be able to execute actions for which he/she has permission". Security policies and specifications incorporate (Includes) security requirements; these may either be security software requirements (SubClass), which relate to the security of applications or databases, or security organizational requirements (SubClass), which relate to assets, persons, or buildings.

3)   Attributes and Axioms of the Security Ontology

In addition to concepts and relationships, an ontology contains axioms and attributes. Formal axioms are assertions accepted as true about abstractions of a field. The axioms allow us to define the meaning of concepts, put restrictions on the values of attributes, examine the conformity of specified information, or derive new concepts [30]. As stated before, the ontology proposed in this paper was not created from scratch. It was constructed by reusing knowledge of existing security ontologies. In particular, some attributes (see Table 1) of the ontology proposed by [31] were reused. For instance, a person has a phone number (its type is Integer); a requirements document has a version (its type is String).

Table 1. Part of the table of attributes

Concept	Attribute	Value type
Person	Phone number	Integer
Software	Version	String
Requirement Document	Version	String
Password	Minimum length	Varchar

The ontology proposed by [24] was a good source of axioms. Table 2 illustrates some axioms with their descriptions and the related concepts.

Table 2. Part of the table of axioms

Description	Expression	Concepts
A threat can be either intentional or accidental	$\forall x: Threat \Rightarrow$ $Intentional\,Threat\,(x) \vee$ $Natural\,Threat\,(x) \vee$ $Accidental\,Threat\,(x)$	Threat
A requirements document can be either a policy or a specification	$\forall x: Requirements\,Document$ $\Rightarrow Security\,Policy\,(x)$ $\vee Specification\,(x)$	Requirements document Security policy Specification

Fig. 4 in the appendix presents the security ontology proposed in this paper. It includes the three dimensions, including concepts and relationships.

2.5   Implementation of the Ontology

Among the different editors of ontologies (OntoEdit [32], Ontolingua [33] and Protégé [34]). Protégé (version 3.4.8) was chosen since it is an extensible, platform-independent environment for creating, editing, viewing, checking constraints, and extracting ontologies and knowledge bases. Ontologies via Protégé can be developed in a variety of formats. OWL 1.0 (Web Ontology Language) was used for the development of the ontology as recommended by the World Wide Web Consortium (W3C). To test and extract relevant knowledge from the security ontology, SQWRL (Semantic Query-Enhanced Web Rule Language) was used. SQWRL is a SWRL-based (Semantic Web Rule Language) for querying OWL ontologies. The description of SQWRL syntax is beyond the scope of the paper; readers may refer to O'Connor et al. [35] for further details. Some indicative queries are presented later in the next section. Implementing the core security ontology with OWL and Protégé is not enough. The target end-users are requirements engineers who are asked to elicit security requirements for different projects, on which they have a tacit knowledge. The ontology will provide the necessary security knowledge in a formalized and explicit form. It also makes available a set of reusable security requirements. To make it usable even for end users not familiarized with Protégé and SQWRL, an interactive environment based on Eclipse was developed. Fig. 2 illustrates the architecture of the tool. The interactive environment facilitates the exploration of the ontology. It automatically and dynamically generates the necessary. SQWRL queries and rules for obtaining the information related to assets, organization, threats, vulnerabilities, and security requirements. The interactive environment makes it possible to generate a specification (a Word document) that summarizes the result of the analysis. Protégé plays the role of the engine; it is intended to wait for SQWRL queries (it plays a passive role in the communication with the end user). Once a query is received,

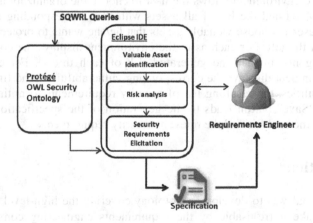

Fig. 2. Architecture of the interactive environment

Protégé processes it and then sends the result to the interactive environment. With this architecture, Protégé is opaque to the requirements engineers; i.e., the requirements engineers do not interact directly with it.

A screenshot of the user interface is presented in Fig. 3. In particular, this figure presents part of the interface; a typical security requirements analysis process was performed, with 3 main windows: valuable asset identification (on the left side), risk analysis, and security requirements elicitation (on the right side).

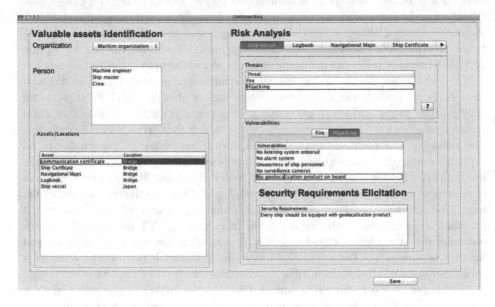

Fig. 3. A screenshot of the interactive environment[5]

The interactive environment allows the user to choose the organization. It displays the persons involved and the list of all assets with their corresponding locations. It also allows the user to choose valuable assets that he/she wants to protect. The latter are displayed on the left. For each asset the environment displays the corresponding threats (threat agents, impact and generated risk of each threat). For each chosen threat, the environment displays the corresponding vulnerabilities. And finally, for all chosen vulnerabilities, the resulting list of security requirements to mitigate them is presented. The "Save" button leads to the generation of the specification document that summarizes the analysis and the relevant security requirements.

3   Evaluation

Given that our goal was to develop an ontology covering the high-level concepts of security, and make it (re)usable by the requirements engineering community, the focus was on the following criteria:

---

[5] A demonstration video can be viewed at: http://youtu.be/zwGbe0Z_mTE

- *Completeness*: this criterion will be evaluated by mapping the target ontology and some other ontologies extracted from literature. The focus was mainly on security ontologies that have been used in security requirements engineering [9][10][11][12].
- *Validity*: Through this criterion, the ability of the ontology to provide reliable answers to a set of questions using its terminology was checked.
- *Usability*: This criterion refers to the "extent to which a product can be used by specified users to achieve specified goals with effectiveness, efficiency and satisfaction in a specified context of use"[6]. In our case, it demonstrates that the ontology can be used for security requirements elicitation, and reused through different projects.

3.1   Completeness

The completeness criterion verifies that our ontology integrates the knowledge that exists in the other ontologies. By completeness, we want to prove that the proposed ontology is 'more' complete than the ones covered by our literature. An alignment table was drawn up, with the concepts of our ontology on one side, and concepts of security ontologies found in security requirements engineering literature on the other side. Table 3 presents the result of the alignment.

Table 3. The alignment table of the proposed security ontology with ontologies used for security requirements elicitation

Concepts of the ontology	Ontologies used for security requirements elicitation				
	Daramola et al. [11]	*Ivankina et al.* [10]	*Lashras et al.* [12]	*Salini et al.* [13]	*Dritsas et al.* [14]
Asset	Asset	Asset	Asset	Asset	Asset
Location	-	-	-	-	-
Organization	-	-	-	-	-
Person	-	-	-	Stakeholder	Stakeholder
Threat	Threat/ Active attack	-	Threat	Threat	Threat/ Deliberate attack
Vulnerability	-	Threat causes	-	Vulnerability	Vulnerability
Risk	-	-	Risk	-	-
Severity	-	-	Valuation criteria	-	-
Impact	-	-	-	Impact severity	-
Threat agent	-	-	-	-	Attacker
Attack tool	-	-	-	-	-
Attack method	Code injection	-	-	-	-
Security goal	-	-	-	-	Objective
Security criterion	-	-	-	Security objective	Security requirement
Security requirement	-	Treatment	Security requirement	Security requirement	-
Control	-	-	Safe guard	-	Countermeasure

---

[6] According to: ISO 9241-11.

Most of the security ontologies used in the SRE contain the concept of "Asset". Given that security issues affect all the infrastructure of organizations, other concepts were introduced (with their corresponding sub-classes): Location, Organization and Person. While many of the other security ontologies take into consideration the concept Threat, most of them neglect the concept Risk generated by a threat, and its Severity. Only the ontology proposed by Dritsas et al. [14] uses the concept of "Attacker". Only the ontology used by Daramola et al. [11] includes the concept of "Attack Method". Our proposed security ontology covers the concept "Objective" used by Dritsas et al. [13]. The concept "Security Criterion", missing in the security ontologies [11], [10] and [12] was used in [13] and [14]. Note that [14] considers as a 'security requirement' what other sources consider a 'security criterion' (availability, confidentiality ...). The concept "Security Requirement" was used in [10], [12] and [13]. These results tend to demonstrate that the proposed security ontology is complete with respect to the union of all the other security ontologies used in security requirements studies, since it incorporates all their concepts.

3.2   Validity

According to Uschold & Gruninger [36], informal and formal questions are one way to evaluate an ontology. The ontology must be able to give reliable answers to these questions using its terminology. The ontology was applied to the maritime domain. For now, the application for domain specific cases is done manually, by instantiating the concepts of the core ontology with domain concepts. Ongoing work is being carried to automatize this instantiation.

This section lists a number of questions that a requirements engineer is likely to encounter during the requirements elicitation phase of a development project. These questions should be regarded as indicative of what the ontology can deal with and reason about. Table 4 summarizes some of these questions. Each of the questions is expressed informally in natural language and formally using SQWRL. The answers to the questions are presented in the last column. These queries guide the requirements engineer during the security requirements elicitation process. The process includes: i) valuable assets identification (what are the assets of the organization? Where are they located? What are the persons involved in the organization?), ii) the risk analysis (what are the threats that threaten the asset? Who leads the attack? What is the attack method used?), and iii) security requirements elicitation (what are the security requirements to mitigate the risk? What are the controls needed to implement those security requirements? What are the security criteria that those requirements fulfill?)

Table 4. Informal and formal questions to the ontology

		Queries	Part of result
Valuable asset identification		What are the organizations in the scope of the project? `Organization(?o)` → `sqwrl:select(?o)`	Maritime organization X, Maritime organization Y
		What are the assets to be protected in the maritime organization X? What is the location of each asset? `Has_Asset(Maritime_organizationX,?a)` • `Has_Location(?a, ?l)` → `sqwrl:select(?a, ?l)`	Ship, Navigation maps located in the bridge
Risk analysis		What are threats that threaten the asset "ship"? `threatens(?T,Ship)` → `sqwrl:select(?T)`	Ship hijacking
		Who is responsible for the threat "Ship hijacking"? `LedBy(Ship_hijacking,?A)`→`sqwrl:select(?A)`	Hijacker
		What is the method used by the hijacker to attract the ship? `Threat(Ship_hijacking)`•`Uses(Hijacker,?M)` → `sqwrl:select (?M)`	Fake distress signal
		What are the impacts of such a threat on the ship? `Implies(Ship_hijacking,?I)`→ `sqwrl:select(?I)`	Theft of provision, Hostage
Security requirements elicitation		What are the security requirements to consider to mitigate the risk? `Exploits (Ship hijacking, V?)` • `mitigated_by(?V, ?R)` →`sqwrl:select(?r)`	**Req1.** Every Ship should be equipped with geolocalization products. **Req2.** Every Ship should be equipped with a listening system on board.

This section has demonstrated how the security ontology could be exploited in the security requirements elicitation phase. This can bring the necessary knowledge to the requirements engineers. This sub-section has illustrated one possible application in the maritime field.

3.3   Usability

To evaluate the usability of the core security ontology, a controlled experiment was performed with end users. The protocol of the experiment was adapted from experimental design and analysis methods [42][43]. In order to obtain a representative group of participants [37], we contacted by mail and phone people from security and requirements engineering communities (laboratories, associations, LinkedIn…). People (industrialists or researchers) not related to the field were intentionally excluded. We used the profile page, and the job position to include/exclude a participant. The day of the experiment, 10 participants were present. The average age was 30 years old. Three participants were certified ISO27000, and three had industrial experience with EBIOS [38] (a well-known French risk assessment method). Four were PhD students working on related subjects. The experiment included a presentation of the security ontology (its main concepts and relations), demonstration of the interactive environment, and a session of manipulation by the participants. At the end of the experiment, participants were asked to fill in a questionnaire[7]. The results extracted from these questionnaires are summed up in Tables 5, 6, 7.

Table 5. Average grading usability

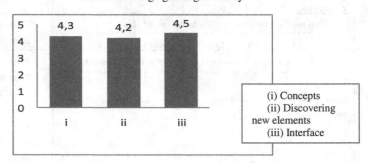

First, the participants were asked to grade the usability of the ontology on a scale of 1 to 5 through three main questions: (i) *Do you find that the security ontology contains the main concepts for security requirements elicitation?* (ii) *Does the security ontology help in finding new elements (threats, vulnerabilities, security requirements, etc.)?* (iii) *Do you find the interface to access to security ontology easy to use?* The scale (1 to 5) corresponds to the degree of agreement to the asked question. Thus (5 = strongly agree, 4 = agree, 3 = neither agree nor disagree, 2= disagree, 1= strongly disagree). Table 5 (page 12) shows a quite high level of satisfaction, which is encouraging. Most participants find that the security ontology includes the main concepts. It helps in discovering new elements even for those who are experts in security since it is not easy to bear in mind hundreds of threats,

---

[7] The questionnaire can be consulted on:
https://www.dropbox.com/s/cc40n31p3fucf4o/Sec%20Ont%20Evaluation%20form.pdf?dl=0

vulnerabilities, and their corresponding security requirements. Almost all participants liked the interactive environment, and revealed that is nice to have the code of the ontology (in OWL-Protégé) hidden. Among the positive qualitative feedbacks that were provided by participants: *"I find in the ontology all concepts that are used in risk analysis methods such as EBIOS"*. One participant mentioned that: *"The ontology seems to have main concepts and individuals, however it would be nice to update it constantly, there are new threats appearing every day!"*. That was an interesting point that could be improved in the future by providing a mechanism to update automatically the individuals of the security ontology.

The second series of questions were particularly devoted to the next stages of the research project and their answers constitute an important input for future work. The participants were asked: (iv) *Does the core security ontology help in building security models?* Secure Tropos models [39] were taken as an example of a security modeling framework. It was presented to participants who did not know it before.

Table 6. Does the core security ontology help in building Secure Tropos models?   **Table 7.** Does the security ontology help in eliciting security requirements for other specific domains?

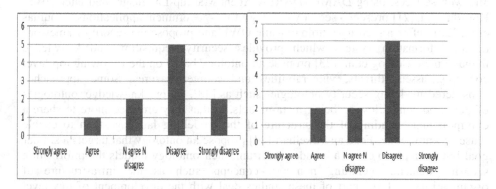

Table 6 reports the results for question (iv). Most participants find it difficult to pass from the concepts of the core security ontology to the concepts of Secure Tropos. A common answer was: *"We understand the existence of connections but the mapping from the core security ontology to Secure Tropos is not straightforward"*. The discussion with participants that followed this question shows that, although the security ontology has the main concepts, relations and individuals, this is still not enough for users to build security models with it. More guidelines or mapping rules are necessary, not for the ontology itself but for the process of using it for security requirements elicitation. The last question was: (v) *does the security ontology help in eliciting security requirements for other specific domains (health, military, and bank)?* We wanted to know if the security ontology helps in providing more security domain specific knowledge each time one switches from a domain to another one. Table 7 reports results for question (v) and shows that most participants "disagree" on the fact that the security ontology by itself is sufficient for eliciting security requirements for different specific domains. One participant mentioned "*something additional is required for the application to different specific domains*". The ontology can be used in different application contexts with some extra collaboration with domain experts, consulting documentation. On the current research phases, we are trying to make the process automatic by using the core security ontology with different domain ontologies.

4   Security Ontologies: Related Works

Considerable works have been devoted to knowledge in the field of security. Schumacher [46] proposed a security ontology and qualifies it as "Core Ontology". This ontology was a good beginning but omits organizational related concepts and some other key concepts such as attack method and attack tool, or security criteria and controls. Undercoffer et al. [17] propose an ontology that characterizes the domain of computer attacks and intrusions. The ontology covers concepts such as host, system component attack, input and consequence. Geneiatakis and Lambrinoudakis [18] propose an ontology for SIP-VoIP (Session Initial Protocol-VoIP) based services. Denker et al. [19][20] develop several ontologies for security annotations of agents and web services, using DAML (DARPA Agent Markup Language) and later OWL. Karyda et al. [21] present a security ontology for e-government applications. Tsoumas et al. [22] define a security ontology using OWL and propose the security framework of an information system which provides security acquisition and knowledge management. Herzog et al. [23] propose an ontology based on the following top-level concepts: assets, threats, vulnerabilities and countermeasures. Some approaches considered modeling security ontologies such as [48]. To our knowledge, ontologies of this kind come close to being meta-models, in that they are used more to share a common understanding of the structure of the modelling language than to enable reuse of knowledge. Fenz and Ekelhart [24] propose an ontology that targets a similar goal but attempts to cover a broader spectrum: their ontology models the information security domain, including non-core concepts such as the infrastructure of organizations. A large part of these studies deal with the development of low-level ontologies limited to a particular domain. A previous survey [7] classifies the existing security ontologies into eight main families: theoretical basis, security taxonomies, general, specific, risk based, web oriented, requirements related and modeling. The analysis of these ontologies reveals that they vary a lot in the way they cover security aspects as reported in previous work [7]. The results converge with those of Blanco et al. who conducted a systematic review of security ontologies [8].

5   Conclusion and Future Work

This paper presents a core ontology for the IS security requirements elicitation and analysis process. The completeness of this ontology was evaluated with regards to existing security ontologies used in security requirements engineering methods. An interactive environment was developed to facilitate its use and reuse. The controlled experiment demonstrated that the ontology helps requirements engineers in eliciting security requirements by allowing them to exploit security-structured knowledge. This was made possible via the interactive environment that dynamically generates the necessary queries. Despite all this effort, the goal of constructing this kind of security ontologies remains ambitious and was found to be more complex than expected. One single team's work is not enough. This research should be of a more collaborative nature including many teams working on security ontologies. A truly complete security ontology remains a utopian goal. However, in the case of this proposed ontology, it can be improved by considering other sources related to security expertise (not mainly ontologies as was the case in this work). The controlled experiment could be performed with a larger number of participants to improve the validity of the results.

In future work, we plan to integrate the ontology and its reasoning features with existing approaches for security requirements analysis (Secure Tropos, KAOS, and others). We plan to make this security ontology more domains-specific by relying on domain ontologies. On the technical level, the plan is to keep the ontology up to date and perform the necessary migrations to the latest available versions (OWL/Protégé).

References

1. Denker, G., Kagal, L., Finin, T.: Security in the Semantic Web using OWL. Information Security Technical Report 10(1), 51–58 (2005)
2. Norton, 2012 Norton Cybercrime report (July 2012)
3. Kauppinen, M., Kujala, S., Aaltio, T., Lehtola, L.: Introducing requirements engineering: how to make a cultural change happen in practice. In: Proceedings IEEE Joint International Conference on Requirements Engineering (RE 2002), pp. 43–51 (2002)
4. Elahi, G., Yu, E., Li, T., Liu, L.: Security Requirements Engineering in the Wild: A Survey of Common Practices. In: Proceedings of COMPSAC 2011, pp. 314–319 (2011)
5. Donner, M.: Toward a Security Ontology. IEEE Security and Privacy 1(3), 6–7 (2003), http://dlib.computer.org/sp/books/sp2003/pdf/j3006.pdf
6. Souag, A.: Towards a new generation of security requirements definition methodology using ontologies. In: Proceedings of 24th International Conference on Advanced Information Systems Engineering (CAiSE 2012), Gdańsk, Poland, June 25-29, pp. 1–8 (2012)
7. Souag, A., Salinesi, C., Comyn-Wattiau, I.: Ontologies for Security Requirements: A Literature Survey and Classification. In: Bajec, M., Eder, J. (eds.) CAiSE Workshops 2012. LNBIP, vol. 112, pp. 61–69. Springer, Heidelberg (2012)
8. Blanco, C., Lasheras, J., Valencia-Garcia, R., Fernandez-Medina, E., Toval, A., Piattini, M.: A Systematic Review and Comparison of Security Ontologies. In: The Third International Conference on Availability, Reliability and Security, ARES 2008, pp. 813–820 (2008)
9. Souag, A., Salinesi, C., Wattiau, I., Mouratidis, H.: Using Security and Domain Ontologies for Security Requirements Analysis. In: IEEE 37th Annual Computer Software and Applications Conference Workshops (COMPSACW), pp. 101–107 (2013)
10. Salinesi, C., Ivankina, E., Angole, W.: Using the RITA Threats Ontology to Guide Requirements Elicitation: an Empirical Experiment in the Banking Sector. In: First International Workshop on Managing Requirements Knowledge, MARK 2008, pp. 11–15 (2008)

11. Daramola, O., Sindre, G., Moser, T.: Ontology-Based Support for Security Requirements Specification Process. In: Herrero, P., Panetto, H., Meersman, R., Dillon, T. (eds.) OTM-WS 2012. LNCS, vol. 7567, pp. 194–206. Springer, Heidelberg (2012)
12. Velasco, J.L., Valencia-Garcia, R., Fernandez-Breis, J.T.,, T.: Modelling Reusable Security Requirements Based on an Ontology Framework. Journal of Research and Practice in Information Technology 41(2), 119 (2009)
13. Salini, P., Kanmani, S.: A Knowledge-oriented Approach to Security Requirements for an E-Voting System. International Journal of Computer Applications 49(11), 21–25 (2012)
14. Dritsas, S., Gymnopoulos, L., Karyda, M., Balopoulos, T., Kokolakis, S., Lambrinoudakis, C., Katsikas, S.: A knowledge-based approach to security requirements for e-health applications. Electronic Journal for E-Commerce Tools and Applications (2006)
15. Massacci, F., Mylopoulos, J., Zannone, N.: An ontology for secure socio-technical systems. Handbook of Ontologies for Business Interactions. IDEA Group (2007)
16. Blanco, C., Lasheras, J., Fernández-Medina, E., Valencia-García, R.,, T.: Basis for an integrated security ontology according to a systematic review of existing proposals. Computer Standards and Interfaces 33(4), 372–388 (2011)
17. Undercoffer, J., Joshi, A., Pinkston, J.: Modeling Computer Attacks: An Ontology for Intrusion Detection. In: The 6th International Symposium on Recent Advances in Intrusion Detection, pp. 113–135 (2003)
18. Geneiatakis, D., Lambrinoudakis, C.: An ontology description for SIP security flaws. Computer Communications 30(6), 1367–1374 (2007)
19. Denker, G., Kagal, L., Finin, T.W., Paolucci, M., Sycara, K.: Security for DAML Web Services: Annotation and Matchmaking. In: Fensel, D., Sycara, K., Mylopoulos, J. (eds.) ISWC 2003. LNCS, vol. 2870, pp. 335–350. Springer, Heidelberg (2003)
20. Denker, G., Nguyen, S., Ton, A.: OWL-S Semantics of Security Web Services: a Case Study. In: Bussler, C.J., Davies, J., Fensel, D., Studer, R. (eds.) ESWS 2004. LNCS, vol. 3053, pp. 240–253. Springer, Heidelberg (2004)
21. Karyda, M., Balopoulos, T., Dritsas, S., Gymnopoulos, L., Kokolakis, S., Lambrinoudakis, C., Gritzalis, S.: An ontology for secure e-government applications. In: The First International Conference on Availability, Reliability and Security, ARES 2006, p. 5 (2006)
22. Tsoumas, B., Gritzalis, D.: Towards an Ontology-based Security Management. In: 20th International Conference on Advanced Information Networking and Applications, AINA 2006, vol. 1, pp. 985–992 (2006)
23. Herzog, A., Shahmehri, N., Duma, C.: An Ontology of Information Security. International Journal of Information Security and Privacy 1(4), 1–23 (2007)
24. Fenz, S., Ekelhart, A.: Formalizing information security knowledge. In: Proceedings of the 4th International Symposium on Information, Computer, and Communications Security, New York, NY, USA, pp. 183–194 (2009)
25. Fernández-López, M., Gómez-Pérez, A., Juristo, N.: METHONTOLOGY: From Ontological Art Towards Ontological Engineering. In: Proceedings of the Ontological Engineering AAAI-97 Spring Symposium Series, Stanford University, EEUU (1997)
26. Jones, D., Bench-capon, T., Visser, P.: Methodologies For Ontology Development. In: Proceedings IT&KNOWS Conference of the 15th IFIP World Computer Congress, pp. 62–75 (1998)
27. Mayer, N.: Model-based Management of Information System Security Risk. Presses universitaires de Namur (2012)
28. Vogel, V.: Information Security Guide,
    https://wiki.internet2.edu/confluence/display/itsg2/Overview+to+the+Guide

29. ISO/IEC 13335-1:2004 Information technology – Security techniques – Management of information and communications technology security – Part 1: Concepts and models for information and communications technology security management (2004)
30. Staab, S., Maedche, A.: Axioms are Objects, too – Ontology Engineering beyond the Modeling of Concepts and Relations. In: Workshop on Applications of Ontologies and Problem-Solving Methods, ECAI 2000, Berlin (2000)
31. Lekhchine, R.: Construction d'une ontologie pour le domaine de la sécurité: application aux agents mobiles (2009)
32. Sure, Y., Angele, J., Staab, S.: OntoEdit: Guiding Ontology Development by Methodology and Inferencing. In: Meersman, R., Tari, Z. (eds.) CoopIS 2002, DOA 2002, and ODBASE 2002. LNCS, vol. 2519, pp. 1205–2011. Springer, Heidelberg (2002)
33. Farquhar, A., Fikes, R., Rice, J.: The Ontolingua Server: a tool for collaborative ontology construction. International Journal of Human Computer Studies 46(6), 707–727 (1997)
34. Horridge, M., Knublauch, H., Rector, A., Stevens, R., Wroe, C.: A Practical Guide To Building OWL Ontologies Using The Protégé-OWL Plugin and CO-ODE Tools Edition 1.0. University of Manchester (2004)
35. O'Connor, M.J., Das, A.K.: SQWRL: A Query Language for OWL. In: OWLED, vol. 529 (2009)
36. Uschold, M., Gruninger, M., Uschold, M., Gruninger, M.: Ontologies: Principles, methods and applications. Knowledge Engineering Review 11, 93–136 (1996)
37. Kitchenham, B.A., Pfleeger, S.L., Pickard, L.M., Jones, P.W., Hoaglin, D.C., El Emam, K., Rosenberg, J.: Preliminary guidelines for empirical research in software engineering. IEEE Transactions Software Engineering 28(8), 721–734 (2002)
38. de la Défense Nationale, S.G.: EBIOS-Expression des Besoins et Identification des Objectifs de Sécurité (2004)
39. Mouratidis, H., Giorgini, P.: Secure Tropos: A Security-Oriented Extension of the Tropos Methodology. International Journal of Software Engineering and Knowledge Engineering 17(02), 285–309 (2007)
40. Kim, A., Luo, J., Kang, M.: Security Ontology for Annotating Resources. In Research Lab, NRL Memorandum Report, p. 51 (2005)
41. Martimiano, A.F.M., Moreira, E.S.: An owl-based security incident ontology. In: Proceedings of the Eighth International Protege Conference, pp. 43–44 (2005)
42. Lawrence, P.S.: Experimental design and analysis in software engineering. Annals of Software Engineering 1(1), 219–253 (1995)
43. Davis, F.D.: Perceived Usefulness, Perceived Ease of Use, and User Acceptance of Information Technology. MIS Quarterly, 319–340 (1989)
44. Norton, 2013 Norton Cybercrime report (July 2013)

Appendix

Fig. 4 in the appendix presents the core security ontology. Table 8 in the appendix was built up for ontology concepts definition. It includes the ontologies used as an entrance point.

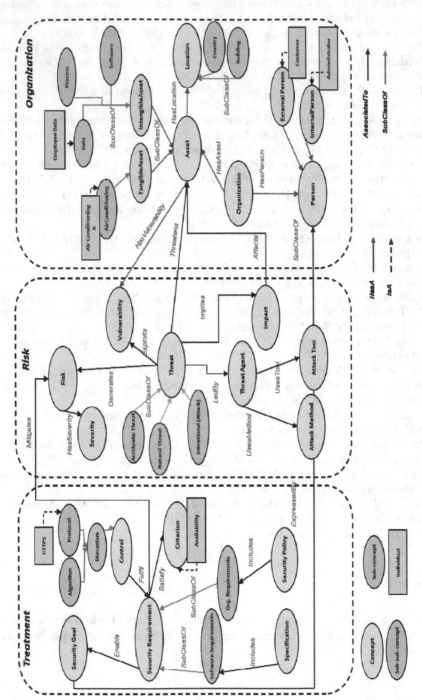

Fig. 4. The core security ontology

Table 8. Ontology concepts definition using security ontologies and models from literature

Dimension	Concepts of the ontology	Mayer et al. [27]	Tsoumas et al. [22]	Herzog et al.[23]	Fenz et al. [24]	Lashera s et al.[12]	Dritsas et al. [14]	Karyda et al.[21]	Kim et al. [40]	Undercoffer et al. [17]	Geneiataki s et al. [18]	Denker et al.[1]	Lekhchine et al.[31]	Martimicano et al. [41]
Organization dimension	Asset	Business Asset	Asset	Asset	Asset	Asset	Asset	Asset	-	System component	Target	-	Asset	Asset
	Location	-	-	-	Location	-	-	-	-	Location	-	-	-	-
	Organization	-	-	-	Organization	-	-	-	-	-	-	-	Organization	-
	Person	-	Stakeholder	-	Person	-	Stakeholder	Person	-	-	-	-	Person	Supplier
Risk dimension	Threat	Threat	Threat	Threat	Threat	Threat	Threat	Threat	-	Means	-	-	Threat	Attack
	Vulnerability	Vulnerability	Vulnerability	Vulnerability	Vulnerability	-	Vulnerability	-	-	-	-	-	Vulnerability	Vulnerability
	Risk	Risk	Risk	-	-	Risk	-	-	-	-	-	-	-	-
	Severity	-	-	-	Attribute	-	-	-	-	-	-	-	-	-
	Impact	Impact	Impact	-	-	Valuation criteria	-	-	-	Consequence	-	-	-	Consequence
	Threat agent	Threat agent	Threat agent	-	Role	-	Attacker	-	-	-	-	-	-	Agent
	Attack method	Attack method	Attack	-	-	-	Deliberate attack	-	-	Intrusion	SIP_attack	-	-	-
	Attack tool	-	-	-	-	-	-	-	-	-	SIP message	-	-	Tool
Treatment dimension	Security goal	-	-	goal	Security attribute	-	Objective	Security objective	Security objective	-	-	-	Objective mechanism	-
	Security criterion	Security criterion	-	-	-	-	-	-	-	-	-	Security notation	-	Acess
	Requirements document	-	Security policy	-	-	-	-	-	-	-	-	-	-	-
	Control	Control	Countermeasure	Countermeasure	Control	Safeguard	Countermeasure	Counter-measure	Security mechanism/ Credential	-	-	Security mechanism/ Credential	Countermeasure/ Protocol/ Algorithm	Correction
	Security requirement	Security requirement	Control	-	Control	Security requirement	-	-	-	-	-	-	-	-

Producing Hook Placements to Enforce Expected Access Control Policies

Divya Muthukumaran[1], Nirupama Talele[2], Trent Jaeger[2], and Gang Tan[3]

[1] Imperial College, London, United Kingdom
[2] The Pennsylvania State University, University Park, PA, United States
[3] Lehigh University, Bethlehem, PA, United States

Abstract. Many security-sensitive programs manage resources on behalf of mutually distrusting clients. To control access to resources, *authorization hooks* are placed before operations on those resources. Manual hook placements by programmers are often incomplete or incorrect, leading to insecure programs. We advocate an approach that automatically identifies the set of locations to place authorization hooks that mediates all security-sensitive operations in order to enforce expected access control policies at deployment. However, one challenge is that programmers often want to minimize the effort of writing such policies. As a result, they may remove authorization hooks that they believe are unnecessary, but they may remove too many hooks, preventing the enforcement of some desirable access control policies.

In this paper, we propose algorithms that automatically compute a minimal authorization hook placement that satisfies constraints that describe desirable access control policies. These *authorization constraints* reduce the space of enforceable access control policies; i.e., those policies that can be enforced given a hook placement that satisfies the constraints. We have built a tool that implements this authorization hook placement method, demonstrating how programmers can produce authorization hooks for real-world programs and leverage policy goal-specific *constraint selectors* to automatically identify many authorization constraints. Our experiments show that our technique reduces manual programmer effort by as much as 58% and produces placements that reduce the amount of policy specification by as much as 30%.

1 Introduction

Programs that manage resources on behalf of mutually distrusting clients such as databases, web servers, middleware, and browsers must have the ability to control access to operations performed when processing client requests. Programmers place *authorization hooks*[1] in their programs to mediate access to such operations[2]. Each authorization hook guards one or more operations and enables the program to decide at runtime whether to perform the operations or not, typically by consulting an access control policy. The access control policy is deployment-specific and restricts the set of operations that each subject is allowed to perform as shown in Figure 1.

---

[1] Authorization Hooks are also known as Policy Enforcement Points.

[2] There are several projects specifically aimed at adding authorization hooks to legacy programs of these kinds [4, 5, 10, 11, 16, 20].

F. Piessens et al. (Eds.): ESSoS 2015, LNCS 8978, pp. 178–195, 2015.
© Springer International Publishing Switzerland 2015

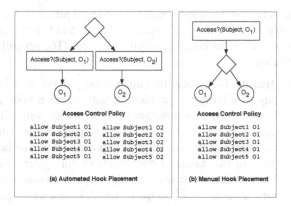

Fig. 1. Figure showing the effect of hook placement granularity on access control policy

There are two main steps programmers perform when placing authorization hooks in programs. First, they must locate the program operations that are security-sensitive. Researchers [7–9, 13, 17] have examined different techniques to infer this using a combination of programmer domain knowledge and automated program analysis. Second, programmers must decide where to place authorization hooks to mediate these security-sensitive operations. Finding location to place authorization hooks is akin to finding *joinpoints* in aspect-oriented programming. Prior automated approaches have suggested hook placements that are very fine grained when compared to placements generated manually by programmers. Understanding this difference is crucial to developing tools for automated hook placement and is the focus of this paper.

The granularity of authorization hook placement affects the granularity of access control policies that must be written for deploying programs. Placing hooks at a very fine granularity means that the access control policy that needs to be written to deploy the program becomes very fine-grained and complex, hindering program deployment. Consider the example (a) in Figure 1 where an automated approach places hooks at two distinct security-sensitive operations O_1 and O_2. If there are five subjects in a deployment, then the policy may need up to 10 rules to specify that the five subjects should have access to both operations.

To facilitate widespread adoption of their programs, programmers want to avoid complex access control policies at deployment. Consequently, they use their domain knowledge to trade off flexibility of the access control policies in favor of simplified policies. For example, an expert may know that when policies are specified for the program, every subject who is authorized to perform O_1 must also be authorized for O_2. With respect to authorization, O_1 and O_2 are equivalent and therefore it is sufficient to insert one hook to mediate both operations as shown in part (b) of Figure 1, resulting in a smaller policy with only five rules.

Minimality in hook placement must be balanced by the two main security requirements of enforcement mechanisms - *complete mediation* [2], which states that every security-sensitive operation must be preceded by an appropriate authorization hook, and the *principle of least privilege*, which states that access control policies only authorize permissions needed for a legitimate prurpose. At present, programmers make these trade-offs between minimality, complete mediation and least privilege manually, which delays the introduction of needed security measures [16] yet still requires many refinements after release.

Authorization Constraints. We believe that making the programmers' domain knowledge explicit is the key to solving the problem of authorization hook placement. We propose an approach that is based on programmers specifying constraints on program operations (e.g., structure member accesses), which we call *authorization constraints*. Authorization constraints restrict the access control policies that may be enforced over a set of operations. We use two kinds of constraints in our system. First, *operation-equivalence* means that two distinct security-sensitive operations on parallel code paths must be authorized for the same set of subjects, which enables a single authorization hook to mediate both. The second kind is *operation-subsumption*, meaning that a subject who is permitted to perform one operation must be allowed to perform the second operation. Therefore, if the second operation follows the first in a code path, then an authorization hook for the second operation is unnecessary.

To use authorization constraints to place authorization hooks automatically, we have to overcome two challenges. First, given a set of authorization constraints, a method needs to be developed to use these constraints to minimize the number of authorization hooks placed to enforce any policy that satisfies the constraints. Second, there should be a method to help programmers discover authorization constraints. In this paper, we provide solutions to both of these challenges.

Minimizing Hook Placement. We address the first challenge by developing an algorithm that uses authorization constraints to eliminate hooks that are unnecessary. Specifically, we show that: (1) when operations on parallel code paths are *equivalent*, the hook placement can be *hoisted*, and (2) when operations that *subsume* each other also occur on the same control-flow path, we can *remove* the redundant hook. This method is shown to produce a minimal-sized[3] hook placement automatically can enforce any access control policy that satisfies the authorization constraints.

Discovering Authorization Constraints. In order to help programmers avoid the cumbersome task of writing complete authorization constraints, we introduce the notion of *constraint selectors*. These selectors are able to make a set of constraint choices on behalf of the programmer based on higher-level goals. For example, suppose the programmer's goal is for her program to enforce multi-level security (MLS) policies, such as those expressed using the Bell-La Padula model [3]. The Bell-La Padula model only reasons about read-like and write-like operations, so any two security-sensitive operations that only perform reads (or writes) of the same object are equivalent. Thus, a constraint selector for MLS guides the method to create equivalence constraints automatically for such operations.

We have designed and implemented a source code analysis tool for producing authorization hook placements that reduce manual decision making while producing placements that are minimial for a given set of authorization constraints. The tool requires

---

[3] Minimality in hook placement is conditional on having precise alias analysis and path sensitive analysis.

only the program source code and a set of security-sensitive operations associated with that code, which can be supplied by one of a variety of prior methods [7–9, 13, 17]. Using the tool, the programmer can choose a combination of constraint selectors, proposed hook placements, and/or hoisting/removal choices to build a set of authorization constraints. Once constraints are established, the tool computes a complete authorization hook placement that has minimum number of hooks with respect to the constraints. We find that using our tool reduces programmer effort by reducing the number of possible placement decisions they must consider, by as much as 67%. Importantly, this method removes many unnecessary hooks from fine-grained, default placements. For example, simply using the MLS constraint selector removes 124 hooks from the default fine-grained placement for X Server.

In this paper, we demonstrate the following contributions:

- We introduce an automated method to produce a minimal authorization hook placement that can enforce any access control policy that satisfies a set of authorization constraints.
- We simplify the task of eliciting authorization constraints from programmers, by allowing them to specify simple high level security goals that are translated into authorization constraints automatically.
- We evaluate a static source code analysis that implements the above methods on multiple programs, demonstrating that this reduces the search space for programmers by as much as 58% and produces placements that reduce the number of hooks by as much as 30%

We believe this is the first work that utilizes authorization constraints to reduce programmer effort to produce effective authorization hook placements in legacy code.

2 Motivation

2.1 Background on Hook Placement

Authorization is the process of determining whether a *subject* (e.g., user) is allowed to perform an *operation* (e.g., read or write an object)[4] by checking whether the subject has a *permission* (e.g., subject and operation) in the *access control policy* that grants the subject with access to the operation. Authorization is necessary because some operations are *security-sensitive operations*, i.e., operations that cannot be performed by all subjects, so access to such operations must be controlled. An *authorization hook* is a program statement that submits a query to check for an authorizing permission in the access control policy. The program statements guarded by the authorization hook may only be executed following an authorized request (control-flow dominance). Otherwise, the authorization hook will cause some remedial action to take place.

Related Work: The two main steps in the placement of authorization hooks are the identification of security-sensitive operations and the identification of locations in the

---

[4] We acknowledge that many access control policies distinguish objects (e.g., files) from the accesses (e.g., read or write), which are often called operations. We define operations to include the object and access type in this paper.

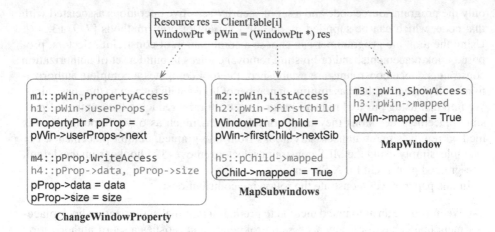

Fig. 2. Hook Placement for functions MAPWINDOW, MAPSUBWINDOWS, and CHANGE
WINDOWPROPERTY

code where authorization hooks must be placed in order to mediate such operations. Past
efforts focused mainly on the first problem, defining techniques to identify security-
sensitive operations [6–9, 13, 15, 17–19, 23]. Initially, such techniques required programmers to specify code patterns and/or security-sensitive data structures manually,
which are complex and error-prone tasks. However, over time the amount of information that programmers must specify has been reduced. In our prior work, we infer
security-sensitive operations only using the sources of untrusted inputs and language-
specific lookup functions [13].

When it comes to the placement of authorization hooks, prior efforts typically suggest placing a hook before every security-sensitive operation in order to ensure complete mediation. There are two problems with this simple approach. First, automated
techniques often use low-level representations of security-sensitive operations, such as
individual structure member accesses, which might result in many hooks scattered
throughout the program. More authorization hooks mean more work for programmers
in maintaining authorization hooks and updating them when security requirements
change. Second, such placements might lead to redundant authorization, as one hook
may already perform the same authorization as another hook that it dominates. In our
prior work, we have suggested techniques to remove hooks that authorize structure
member accesses redundantly [13]. However, this approach still does not result in a
placement that has a one-to-one correspondence with hooks placed manually by domain experts. In X server, it was found that while the experts had placed approximately
200 hooks, the automated technique suggested approximately 500 hooks. In the following subsections we discuss some reasons for this discrepancy.

2.2 How Manual Placements Differ

We find that there are typically two kinds of optimizations that domain experts perform
during hook placement. We follow with examples of both cases from the X Server.

- First, assume there are two automatically placed hooks H_1 and H_2 such that the former dominates the latter in the program's control flow. The placement by the domain expert has a matching hook for H_1 but not for H_2. We can interpret this as the expert having *removed* (or otherwise omitted) a finer-grained hook because the access check performed by H_1 makes H_2 redundant.
- The automated tools place hooks $(H_1, ..., H_n)$ at the branches of a control statement. The domain expert has not placed hooks that map to any of these hooks, but instead, has placed a hook M_1 that dominates all of these hooks[5]. The expert has *hoisted* the mediation of operations at the branches to a common mediation point above the control statement as shown in Figure 1.

First, examine the code snippet in Figure 2. In the figure, hooks placed by the programmer have prefixes such as m1:: and hooks placed by an automated tool [13] have prefixes such as h1::. The function MAPWINDOW performs the operation write(pWin→mapped) on the window, which makes the window viewable. We see that the programmer has placed a hook m3::(pWin,ShowAccess) that specifically authorizes a subject to perform this operation on object represented by pWin. Access mode ShowAccess identifies the operation. The requirement of consistency in hook placement dictates that an instance of hook ShowAccess should precede any instance of writing to the mapped field of a Window object that is security-sensitive. MAPSUBWINDOWS performs the same operation on the child windows pChild of a window pWin. While the automated tool prescribes a hook at MAPSUBWINDOWS, we find that the domain expert has chosen not to place a corresponding hook. MAPSUBWINDOWS is preceded by the manual hook m2::(pWin,ListAccess) for the subject to be able to list the child windows of window pWin, but there is no hook to authorize the operation write(pChild->mapped)[6].

Second, look at the example shown in Listing 1.1. The function COPYGC in the X server accepts a source and target object of type GC and a mask that is determined by the user request and, depending on the mask, one of 23 different fields of the source are copied to the target via a switch-case control statement. Since each branch results in a different set of structure member accesses, the automated tool infers that each branch performs a different security-sensitive operation. Therefore, it suggests placing a different hook at each of the branches. On the contrary, there is a single manually placed hook that dominates the control statement, which checks if the client has the permission DixGetAttrAccess on the source object. Therefore a single manually placed hook replaces 23 automated hooks in this example.

2.3 Balancing Minimality with Least Privilege

We have in the X Server a mature codebase, which has been examined over several years by programmers in order to reach a consensus on hook placement. We are convinced that a deliberate choice being made by the experts about where to place hooks and which hooks to avoid on a case-by-case basis. For example, in CHANGEWINDOWPROPERTY in Figure 2, a property pProp of pWin is retrieved and accessed. Programmers have

---

[5] There is no hook in the automated placement that matches M_1.
[6] We discuss the relevance of CHANGEWINDOWPROPERTY below.

Listing 1.1. Example of manual hoisting in the COPYGC function in the X server

```
/*** gc.c ***/
int CopyGC(GC *pgcSrc, GC *pgcDst, BITS32 mask){
 switch (index2)
   {
        result = dixLookupGC(&pGC, stuff->srcGC,
            client,DixGetAttrAccess);
        if (result != Success)
            return BadMatch;
        case GCFunction:
            /* Hook(pgcSrc, [read(GC->alu)]) */
            pgcDst->alu = pgcSrc->alu;
            break;
        case GCPlaneMask:
            /* Hook(pgcSrc, [read(GC->planemask)]) */
            pgcDst->planemask = pgcSrc->planemask;
            break;
        case GCForeground:
            /* Hook(pgcSrc, [read(GC->fgPixel)]) */
            pgcDst->fgPixel = pgcSrc->fgPixel;
            break;
        case GCBackground:
            /* Hook(pgcSrc, [read(GC->bgPixel)]) */
            pgcDst->bgPixel = pgcSrc->bgPixel;
            break;
        /* .... More similar cases */
    }
}
```

placed a finer-grained hook m4::(pProp,WriteAccess) in addition to the hook m1::(pWin,SetPropertyAccess). Contrast this with the MAPSUBWINDOWS example where they decided not to mediate the access of a child object.

The fundamental difference between a manually written hook placement and an automatically generated one is in the granularity at which security-sensitive operations are defined. When the automated tool chooses to place a hook at each of the branches of a control statement, it implicitly identifies security-sensitive operations at a finer granularity than experts. The choice of granularity of security-sensitive operations is an exercise in balancing the number of hooks placed and least privilege. A fine-grained placement allows more precision in controlling what a subject can do, but this granularity may be overkill if programmers decide that subjects must be authorized to access operations in an all-or-nothing fashion. For the switch statement with 23 branches, having 23 separate hooks will lead to a cumbersome policy because the policy will have 23 separate entries for each subject. Since the programmers decided that all subjects either can perform all 23 operations for an object or none, it is preferable to have a single hook to mediate the 23 branches.

We have also seen that even with manual hook placement multiple iterations may be necessary to settle on the granularity that balances least privilege and minimality in hook placement. For example, the X server version of 2007 had only four operation modes, namely, read, write, create and destroy. But during the subsequent release, the programmers replaced these with 28 access modes that were necessary to specify and enforce policies with finer granularity. Since the first release of X server with the XACE hooks in 2007, the hooks have undergone several changes. Over 30 hooks were added to the X server code base, and some existing hooks were also removed, moved or combined with other hooks [22]. Some of these changes are documented in the XACE specification [21]. We believe that observing typical policy specifications at runtime enabled the programmers to add and remove hooks in subsequent versions of the application. We want to understand iterative refinement and build methods to automate some tasks in the process, making programmer decisions explicit.

3 Authorization Hook Placement Problem

Authorization hook placement involves two main steps: a) finding security-sensitive operations (SSOs) in the program and, b) placement of hooks to satisfy a set of requirements. In this section we give a brief background into approaches that tackle the former, followed by our intuition for how to approach the latter problem.

3.1 Identifying Security-Sensitive Operations

We provide some background on our prior research [13] in automatically identifying the set O of security-sensitive operations (SSOs) in programs using static analysis. Each SSO is represented using a variable v and a set of read and write structure member accesses on the variable. For example, in the CHANGEWINDOWPROPERTY function shown in Figure 2, the last two statements produce a security-sensitive operation pProp: [write(data), write(size)]. There may be multiple instances of an SSO in a program. Each instance is represented using the tuple (o, l) where o is the SSO and l is the location (statement) in the code where the instance occurs. Let O_L be the set of all instances of all the SSOs in the program. Our goal is to place authorization hooks to mediate all the elements of O_L.

Definition 1. *An authorization hook is a tuple (O_h, l_h) where l_h is a statement that contains the hook and $O_h \subseteq O_L$ is a set of security-sensitive operation instances mediated by the hook.*

A set of authorization hooks is called an *Authorization Hook Placement*. The approach in [13] produces a Control Dependence Graph (CDG) of the program to represent program statements and hooks. A CDG of a program $CDG = (L, E)$ consists of a set of program statements L and edges E, which represent the control-dependence relationships between statements. Since this exposes the statements that a given statement depends upon for execution, it enables computation of authorization hook placements that mediate all control flows in the program by ensuring that every operating instance in O_L is included in at least one authorization hooks O_h. We will continue to use the CDG representation in this work for refining hook placements.

3.2 Consolidating Hook Placements

As mentioned in Section 2, the inital placement of hooks by automated techniques is fine-grained, i.e., typically at every security-sensitive operation instance that is identified. Our intuition is that constraints on the access control policies to be enforced in the program can be leveraged to consolidate authorization hooks in order to achieve the right granularity for hook placements. Let U be the set of all subjects for a hypothetical access control policy $\mathcal{M}$ for the given program. Let *Allowed* be a function that maps each security-sensitive operation in O to subjects U that are allowed to perform the operation according to policy $\mathcal{M}$. We identify two cases that are relevant to the placement of authorization hooks:

- **Invariant I:** First, given any two operations o_1 and o_2, if access control policy $\mathcal{M}$ permits $Allowed(o_1) = Allowed(o_2)$ then o_1 and o_2 are equivalent for the purpose of authorization. This means that any hook that mediates o_1 and dominates o_2 in the code automatically authorizes o_2 and vice versa.
- **Invariant II:** Second, given two operations o_1 and o_2, if access control policy $\mathcal{M}$ permits $Allowed(o_1) \subset Allowed(o_2)$ then operation o_1 'subsumes' o_2 for the purpose of authorization. This means that a hook that mediates o_1 and dominates o_2 can also mediate o_2 but not vice-versa.

As described above, we have observed that programmers often assume that their programs will enforce access control policies that satisfy Invariant I (e.g., see Listing 1.1) and Invariant II for *every* access control policy $\mathcal{M}$ that may be enforced by their programs (see Figure 2). This observation leads us to believe that in order to consolidate hooks we need to impose equivalence and partial-order relationships between the elements in O. Therefore, we define a *set of authorization constraints* as follows:

Definition 2. *A set of authorization constraints $\mathcal{P}$ is a pair (S, Q) of relationships between SSOs in the program, where Q stands for* **equivalence** *and S stands for* **subsumption**.

We can see that the equivalence relationship Q results in a partitioning of the set O of security-sensitive operations. Let O_Q be the set of partitions produced by Q. The subsumption relationship S imposes a partial order between the elements in O_Q. We can use these two relationships to consolidate hooks on parallel code paths to a dominating program location (hook *hoisting* operation) or eliminate a redundant hook on a post-dominating program location (hook *removal* operation). We describe the technique for this in the Section 4.

The challenge we address in this paper is how to use authorization constraints to consolidate hook placements. Our system uses an algorithm that, given a program and its set of authorization constraints, generates a minimal authorization hook placement that satisfies complete mediation and least privilege. **Complete mediation** states that every SSO instance in a program should be dominated by an authorization hook that mediates access to it. A placement that enforces **least privilege** ensures that during an execution of a program, a user of the program (subject) is only authorized to perform (or denied from performing) the SSOs requested as part of that execution. This effectively puts a constraint on how high a hook can be hoisted.

We observe that even though automated placement methods may be capable of producing placements that can enforce any policy, and thus can enforce least privilege, programmers will not accept a hook placement unless there is a justified need for that hook. Specifically, programmers only want a hook placed if that hook will actually be necessary to prevent at least one subject from accessing a security-sensitive operation. That is, while programmers agree that they should give subjects the minimal rights, they also require *that a program should have only the minimal number of hooks necessary to enforce complete mediation and least privilege.*

4   Design

There are two main inputs to our approach: a) the set O of SSOs and their instances O_L in the program, and b) the control dependence graph CDG of the program. We have defined these inputs in Section 3.1 and discussed prior work that uses static analysis to infer them automatically.

Step 1: Generating a Default Placement: First, using O, O_L and CDG, we generate a default placement. Since the elements in O_L are in terms of code locations, and the nodes in the CDG have location information, we can create a map $C2O$ from each node s_i in the CDG to the set of SSO instances in O_L that occur in the same location as s_i. The default placement $\mathcal{D}$ has a hook at each node s_i where $C2O[s_i] \neq \emptyset$.

Step 2: Generating Constrained Hook Placement: If a set of authorization constraints $\mathcal{P}_i$ is already available, step 2 of our tool is able to automatically generate a minimal placement[7] $\mathcal{E}_i$ that can enforce any access control policy that satisfies the constraints. Our approach uses the *equivalence* constraints to perform *hoisting* and *subsumption* constraints to perform *removal*, thereby minimizing the number of hooks. This procedure is described in Section 4.1. We discuss how we may assist programmers in selecting authorization constraints for their program using high-level goals encoded as constraint selectors in Section 4.2.

4.1   Deriving a Constrained Placement

Given default placement $\mathcal{D}$ and set of authorization constraints $\mathcal{P}$, our system can derive a candidate constrained placement $\mathcal{E}$ that satisfies complete mediation and least privilege enforcement with the minimum number of hooks (used in step 3 of our design). The subsumption S and equivalence Q relationships in $\mathcal{P}$ enable us to perform two different hook refinements on $\mathcal{D}$ to derive the $\mathcal{E}$. We present the algorithm next and the proof why it has desired properties is in Appendix A.

Hoisting. The first refinement is called *hoisting* and it aims to consolidate the hooks for mediation of equivalent operation instances that are siblings (appear on all branches of control statements). This lifts hook placements higher up in the CDG based on Q. Given a node s_i in the CDG, if each path originating from that node s_i contains SSOs that are in the same equivalence class in O_Q, then we can replace the hooks at each of these paths with a single hook at s_i. This relates to the example in Listing 1.1, where if the operations along all the 23 branches of the $Switch$ statement are equivalent, then we can replace the 23 automatically generated hooks at those branches with a single hook that dominates all of them.

Algorithm 1 shows how hoisting is done. It uses the CDG and the $C2O$ map as inputs. Accumulator α gathers the set of SSOs at each node s_i by combining the $C2O$ of s_i with the α mapping from the child nodes. The algorithm traverses the CDG in reverse topological sort order and makes hoisting decisions at each node in the CDG. We partition the set of nodes in the CDG into two types - control and non-control nodes. Control nodes represent control statements (such as if, switch etc) where hoisting can

---

[7] Henceforth referred to as a *constrained placement*.

Algorithm 1. Algorithm for hoisting

$top' = TopoSortRev(CDG)$
while $top' \neq \emptyset$ **do**
    $s_i = top'.pop()$
    if $isControl(s_i)$ **then**
        $\alpha[s_i] = C2O[s_i] \cup \bigcap_j^Q \{\alpha[s_j] \mid (s_i, s_j) \in CDG\}$
    else
        $\alpha[s_i] = C2O[s_i] \cup \bigcup_j \{\alpha[s_j] \mid (s_i, s_j) \in CDG\}$
    end if
end while

Algorithm 2. Algorithm for removal

$top = TopoSort(CDG)$
while $top \neq \emptyset$ **do**
    $s_i = top.pop()$
    $O_D = \bigcap_j^{QS} \{\phi[s_j] \mid (s_j, s_i) \in CDG\}$
    $\phi[s_i] = \alpha[s_i] \cup O_D$
    $O_R = \emptyset$
    for all $o_m \in \alpha[s_i]$ **do**
        if $\exists o_n \in O_D, (o_n \ S \ o_m)$ or $(o_n \ Q \ o_m)$ **then**
            $O_R = O_R \cup \{o_m\}$
        end if
    end for
    $\beta[s_i] = \alpha[s_i] - O_R$
end while

be performed. At control nodes[8], we perform the intersection operation $\bigcap^Q$ which uses the equivalence relation Q to perform set intersection in order to consolidate equivalent SSOs. Note that this intersection operation limits how high hooks may be hoisted in the program. At non-control nodes, we accumulate SSOs from children using a union operation.

Note that this algorithm does not remove any hooks. It places new hooks that dominate the control statements where hoisting occurs. For example, given Listing 1.1, the algorithm would place a new hook before the Switch statement. The removal operation which we discuss next will eliminate the 23 hooks along the different branches because of the new hook that was placed by this algorithm.

Removal. The second refinement is called redundancy removal and aims to eliminate superfluous hooks from CDG using S. Whenever a node s_1 that performs SSO o_1 dominates node s_2 that performs SSO o_2 and o_1 either subsumes or is equivalent to o_2 according to $\mathcal{P}$, then a hook at s_1 for o_1 automatically checks permissions necessary to permit o_2 at s_2. Therefore, we may safely remove the hook at s_2 without violating complete mediation.

In the example in Figure 2, if we had authorization constraints specify that operation $read(pWin \rightarrow firstChild)$ subsumes (or is equivalent to) operation $write(pChild \rightarrow mapped)$, then we do not need the suggested hook h5.

Algorithm 2 shows how the removal operation is performed. The algorithm takes as input the CDG and the map α computed by the hoisting phase. It traverses the CDG in topological sort order (top-down) and at each node s_i makes a removal decision based on the set of operations checked by all hooks that dominate s_i. The accumulator ϕ

---

[8] Each control node has dummy nodes as children each representing a branch of the control node.

stores for each node s_i the set of operations checked at s_i and all nodes that dominate s_i. While processing each node, the algorithm computes O_D, which is the set of operations checked at dominators to node s_i. Note that the CDG is constructed interprocedurally (refer to Section 3.1) and a node can have multiple parents at function calls[9]. The $\bigcap^{QS}$ that combines authorized operations in case of multiple parents is shown in Algorithm 3.

Algorithm 3. Compute $\bigcap^{QS}$ on two sets O_i and O_j. Returns result in O_T.

$O_T = \emptyset$
for all $o_i \in O_i$ **do**
   for all $o_j \in O_j$ **do**
      if o_i Q o_j **then**
         $O_T = O_T \cup \{o_i\}$
      end if
      if o_i S o_j **then**
         $O_T = O_T \cup \{o_j\}$
      end if
      if o_j S o_i **then**
         $O_T = O_T \cup \{o_i\}$
      end if
   end for
end for

Next, Algorithm 2 creates the set O_R which is the set of operations that do not have to be mediated at s_i since they are either subsumed by or equivalent to operations that have been mediated at dominators to s_i. The resulting map β from nodes to the set of operations that need to be mediated at the node gives the final placement. The constrained placement $\mathcal{E}$ suggests a hook at each node s_i such that $\beta[s_i] \neq \emptyset$.

Note that both the bottom-up hoisting and top-down removal must be performed in sequence to get the final mapping from nodes to the set of SSOs that need mediation.

4.2 Helping Programmers Produce Placements

In this section we discuss how a programmer might produce a hook placement using the approach presented above. More specifically, would the programmer have to manually generate a fine-grained set of authorization constraints in order to use our approach? We envision an approach where programmers only need to specify high-level security goals as opposed to fine-grained set of authorization constraints.

Suppose the programmer wishes to enforce a well-known security policy, such as Multi-Level Security [12]. In MLS, subjects are assigned permissions at the granularity of read and write accesses to individual objects. In our method, program objects are referenced by variables in program operations, so any MLS policy that permits a subject to read a field of a variable also permits that subject to read any other fields of that variable; a similar case holds for writes. This means that all read-like (write-like) accesses of a variable can be treated as equivalent. The programmer can produce a small

---

[9] We avoid cycles in the CDG by eliminating back edges. This is for the purpose of being able to sort nodes topologically in the CDG. When we perform hook hoisting and removal, it is the control dominance information in a CDG that gets used in our algorithms; so eliminating back edges would not affect our analysis.

Table 1. Table showing the lines of code (LOC), number of manual hooks (MANUAL), default automated hooks (DEFAULT), and the impact of using the MLS constraint selectors for hook placement, including the resultant number of hooks (Total), percent reduction in total hooks (%-Reduction), and the percent reduction in the difference between the manual and automated placements (%-Difference)

Program	LOC	MANUAL	DEFAULT	MLS		
				Total	*%-Reduction*	*%-Difference*
X Server 1.13	28K	207	420	296	30	58
Postgres 9.1.9	49K	243	360	326	9	29
Linux VFS 2.6.38.8	40K	55	139	135	2	5
Memcached	8K	0	32	30	6	n/a

set of such rules to encode the security policy (we call this a constraint selector) which serves as an input to the placement approach in lieu of a complete set of authorization constraints.

Whenever a hoisting or a removal decision needs to be made by the approach, the constraint selector will serve as an oracle that aids in this decision. The MLS constraint selectors will stipulate that whenever all the branches perform either only reads or only writes of the same variable (irrespective of the fields being read) a hoisting operation should succeed. Similarly, if a hook mediated the read operation of a variable and is dominated by a hook that mediates the read of the same variable (irrespective of the field being accessed) the removal operation should succeed. The programmers can create any such constraint selector that encodes relationships between data types that may be application specific, or even encode the results or auxiliary static and dynamic analysis.

5   Evaluation

We implemented our approach using the CIL [14] program analysis framework and all our code is written in OCaml. The CDG construction and default hook placement is similar to the approach mentioned in prior work [13]. Our prototype implementation does not employ precise alias analysis or a path sensitive analysis which may produce some redundant hooks (e.g. in code paths that are not feasible).

Our goal with the evaluation was to answer two questions:
a) Does the approach produce placements that are closer to manually placed hooks?
b) Does the approach reduce programmer effort necessary to place authorization hooks?

In order to evaluate our approach, we compare hook placement produced by using constraint selectors against the default hook placement produced using the technique presented in our prior work [13]. We perform this comparison along two dimensions:

- First, we determine the number of hooks produced using both techniques and show that using constraint selectors reduces the number of hooks by as much as 30%. We also show that using the constraint selectors reduces the gap between manual and automated placements by as much as 58% compared to the default approach [13].
- Second, we show that using constraint selectors reduces the programmer effort, measured in terms of the number of authorization constraint options to manually consider by as much as 58%.

Table 2. Table showing the hoisting (HOIST) and removal (REMOVE) suggestions in the default placement (DEFAULT) and placements generated using the constraint selectors (MLS)

Program	DEFAULT		MLS	
	REMOVE	HOIST	REMOVE	HOIST
X Server 1.13	237	55	113	10
Postgres 9.1.9	208	42	146	21
Linux VFS 2.6.38.8	53	4	49	3
Memcached	8	1	6	0

Evaluating Hook Reduction: Table 1 shows the total number of hooks placed for each experiment. 'LOC' shows the number of lines of code that were analyzed, 'MANUAL' shows the number of hooks placed manually by domain experts, 'DEFAULT' shows the number of hooks placement in the default case by the automated technique and 'MLS' shows how using the constraint selectors affects the total number of hooks and how this compares with the number of hooks placed manually. Within the 'MLS' column, 'Total' refers to the total number of hooks placed when using the constraint selectors '%-Reduction' refers to the percentage reduction in number of hooks compared to the default placement and '%-Difference' refers to percentage reduction in the gap between automated and manual placement when compared against the default placement. For example, in the case of X Server we see that in the experiment we considered 28K lines of code, where programmers had placed 207 hooks manually, whereas the default placement suggested 420 hooks. When the 'MLS' constraint selectors was used, the number of hooks suggested by the automated technique went down to 296 which is a 30% reduction in the number of hooks compared to manual placement and reduces the gap between automated and manual hook placements by 58%.

Evaluating Programmer Effort: We define programmer effort as the search space of authorization constraints that the programmer has to examine manually. Therefore, we measure the reduction in programmer effort by counting the number of *hoisting* and *removal* choices that the tool automatically makes using the constraint selectors after the default placement. The results for this experiment are shown in Table 2 with the number of removal choices (REMOVE) and hoisting choices (HOIST) for each experiment. For example, there were 237 removal choices and 55 hoisting choices available to the programmer after the default placement for X Server. After applying the proposed placement approach using the MLS constraint selectors, there are 113 removal and 10 hoisting choices remaining from which the programmer has to select from. This implies that using the MLS constraint selectors has reduced the number of choices that the programmer has to make to produce their desired placement. Making some set of hoisting and removal choices may expose additional choices due to newly introduced dominance and branch relationships. Therefore the number of choices shown in this table is not a measure of the total remaining programmer effort but only of the next set of choices available to the programmer.

We ran our experiments on four programs:

X Server 1.13. Our results show that we are able to reduce the amount of programmer effort by 58%. The number of hooks generated is reduced by 30% (reducing the gap between manual and automated placements by 58% as well).

Postgres 9.1.9. This version has mandatory access control hooks [1] hooks, but these are incomplete according to the documentation of the module [1]. Therefore, we only consider discretionary access control hooks. Our experiments show that we are able to reduce the amount of programmer effort by 32%. The number of hooks generated is reduced by 15% (reducing the gap between manual and automated placements by 46%).

Linux kernel 2.6.38.8 Virtual File System (VFS). The VFS allows clients to access different file systems in a uniform manner. The Linux VFS has been retrofitted with mandatory access control hooks in addition to the discretionary hooks. Our results show that there is an 17% reduction in programmer effort and four fewer hooks then default placement.

Memcached. This general-purpose distributed memory caching system does not currently have any hooks. Our experiments show that constraint selectors are able to reduce the amount of programmer effort by 33% and the number of hooks by 22%.

In the case of the Linux VFS, 'MLS' does not make a significant dent on the removal choices. All the removal choices in Linux VFS fell into one of three categories. First, 24 of the 49 remaining choices are interprocedural hook dominance relationships where the security-sensitive objects being guarded by the hooks were of different data types. For example, the hook in function do_rmdir dominates the hook in vfs_rmdir, therefore there is a removal opportunity. But the hook in the former mediates an object of type nameidata and the latter mediates an object of type struct dentry *. Our approach currently only performs removal when the hooks mediate the same variable. Second, 19 choices are interprocedural hook dominance relationships where the object mediated is of the same type but because it is across procedure boundaries and our approach does not employ alias analysis, it conservatively assumes that they are different objects. Finally, six choices were due to intraprocedural hooks on the same object but one mediates reads and the other mediates writes. The 'MLS' constraint selector forbids removal in these cases.

6   Conclusion

In this paper we have successfully demonstrated that our automated system can generate minimal authorization hook placements that satisfy complete mediation and least privilege guided by authorization constraints. We show that using static and dynamic analysis techniques to help programmers select these authorization constraints. We show how our technique can be practically used by programmers to reduce the manual effort in weighing different authorization hook placement options.

Acknowledgement. We thank the anonymous reviewers of this paper and the shepherd Dr. Christian Hammer for their valuable feedback in preparing this manuscript. This material is based upon work supported by the National Science Foundation Grant No. CNS-1408880 and in part by DARPA under agreement number N66001-13-2-4040. The U.S. Government is authorized to reproduce and distribute reprints for Governmental purposes notwithstanding any copyright notation thereon.

References

1. F.38. sepgsql (2013),
   http://www.postgresql.org/docs/9.1/static/sepgsql.html
2. Anderson, J.P.: Computer security technology planning study, volume II. Technical Report ESD-TR-73-51, HQ Electronics Systems Division (AFSC) (October 1972)
3. Bell, D.E., LaPadula, L.J.: Secure computer system: Unified exposition and Multics interpretation. Technical Report ESD-TR-75-306, HQ Electronic Systems Division (AFSC) (March 1976)
4. Carter, J.: Using GConf as an Example of How to Create an Userspace Object Manager. In: 2007 SELinux Symposium (2007)
5. Walsh, D.: Selinux/apache,
   http://fedoraproject.org/wiki/SELinux/apache
6. Edwards, A., Jaeger, T., Zhang, X.: Runtime verification of authorization hook placement for the Linux security modules framework. In: Proceedings of the 9th ACM Conference on Computer and Communications Security, pp. 225–234 (2002)
7. Ganapathy, V., Jaeger, T., Jha, S.: Automatic placement of authorization hooks in the Linux Security Modules framework. In: Proceedings of the 12th ACM Conference on Computer and Communications Security, pp. 330–339 (November 2005)
8. Ganapathy, V., Jaeger, T., Jha, S.: Retrofitting legacy code for authorization policy enforcement. In: Proceedings of the 2006 IEEE Symposium on Security and Privacy, pp. 214–229 (May 2006)
9. Ganapathy, V., King, D., Jaeger, T., Jha, S.: Mining security-sensitive operations in legacy code using concept analysis. In: Proceedings of the 29th International Conference on Software Engineering (ICSE) (May 2007)
10. Gong, L., Schemers, R.: Implementing protection domains in the java^{tm} development kit 1.2. In: NDSS (1998)
11. Love, R.: Get on the D-BUS (January 2005),
    http://www.linuxjournal.com/article/7744
12. Multilevel security in the department of defense: The basics (1995),
    http://nsi.org/Library/Compsec/sec0.html
13. Muthukumaran, D., Jaeger, T., Ganapathy, V.: Leveraging "choice" to automate authorization hook placement. In: CCS 2012: Proceedings of the 19th ACM Conference on Computer and Communications Security, page TBD. ACM Press (October 2012)
14. Necula, G.C., McPeak, S., Rahul, S.P., Weimer, W.: Cil: Intermediate language and tools for analysis and transformation of c programs. In: Nigel Horspool, R. (ed.) CC 2002. LNCS, vol. 2304, pp. 213–228. Springer, Heidelberg (2002)
15. Politz, J.G., Eliopoulos, S.A., Guha, A., Krishnamurthi, S.: Adsafety: type-based verification of javascript sandboxing. In: Proceedings of the 20th USENIX Conference on Security, SEC 2011, p. 12. USENIX Association (2011)
16. SE-PostgreSQL? (2009),
    http://archives.postgresql.org/message-id/20090718160600.
    GE5172@fetter.org

17. Son, S., McKinley, K.S., Shmatikov, V.: Rolecast: finding missing security checks when you do not know what checks are. In: Proceedings of the 2011 ACM International Conference on Object Oriented Programming Systems Languages and Applications, OOPSLA 2011, pp. 1069–1084. ACM (2011)
18. Sun, F., Xu, L., Su, Z.: Static detection of access control vulnerabilities in web applications. In: Proceedings of the 20th USENIX Conference on Security, SEC 2011, p. 11. USENIX Association (2011)
19. Tan, L., Zhang, X., Ma, X., Xiong, W., Zhou, Y.: Autoises: automatically inferring security specifications and detecting violations. In: Proceedings of the 17th Conference on Security Symposium, pp. 379–394. USENIX Association (2008)
20. Implement keyboard and event security in X using XACE (2006), https://dev.laptop.org/ticket/260
21. Implement keyboard and event security in X using XACE (2006), https://dev.laptop.org/ticket/260
22. Xorg-Server Announcement (2008), http://lists.x.org/archives/xorg-announce/2008-March/000458.html
23. Zhang, X., Edwards, A., Jaeger, T.: Using CQUAL for static analysis of authorization hook placement. In: Proceedings of the 11th USENIX Security Symposium, pp. 33–48 (August 2002)

A   Hook Placement Properties

First, we want to prove that the authorization hook placement mechanism satisfies two goals: *least privilege* enforcement and *complete mediation*. We start by showing that our initial placement satisfies these properties and the subsequent hoisting and removal phases preserve this property. Least privilege is defined as follows:

Definition 3. *In a least privilege placement a hook (O_h, l_h) placed at location l_h authorizing a set of SSOs O_h implies that for each $o_i \in O_h$, on each path in the program originating from l_h, there must be an operation instance (o_j, l_j) such that $(o_i \, S \, o_j) \vee (o_i \, Q \, o_j)$.*

Complete mediation is defined as:

Definition 4. *Complete Mediation requires that for every operation instance (o_i, l_i), there exists a hook (O_h, l_h) such that l_h control flow dominates l_i and there exists $o_h \in O_h$ such that $(o_i \, Q \, o_h) \vee (o_h \, S \, o_i)$.*

Our approach depends on two inputs a) The set of all SSOs in the program b) The authorization constraints that determine all possible optimizations (hoisting and removal) in hook placement. Our proof assumes that both of these specifications are complete.

Our approach starts by placing a hook at every instance of every SSO in the program. First, it is trivial to show that this results in a placement that adheres to complete mediation since every SSO instance has a corresponding hook. Second, this approach guarantees least privilege since every hook is post-dominated by the SSO instance for which it was placed.

Hoisting. The hoisting in Algorithm 1 hoists the hooks pertaining to equivalent SSOs on all branches of a control statement in a bottom-up fashion in the CDG. The hoisting operation introduces a new hook which dominates the control statement. This new hook preserves least privilege since all the branches of the control statement (therefore all paths originating from the new hook) must contain instances of operations that are equivalent to the one mediated by the new hook. This stage does not remove any hooks so complete mediation is preserved.

Removal. The redundancy removal stage in Algorithm 2 propagates information about hooks placed in a top-down fashion in the CDG. The removal operation does nothing to violate least privilege since it does not add additional hooks. When each node n of the CDG is processed, the set of propagated hooks that reach n represent the hooks that control dominate n. Therefore, if a hook h placed at node n is subsumed by or equivalent to any hook in the set of propagated hooks, then h can be safely removed without violating the complete-mediation guarantee.

Additionally given sound and complete alias analysis we can also guarantee a minimality in hook placement (constrained by complete mediation and least privilege). The construction of our algorithm guarantees that both hoisting and removal at each node are performed transitively in the context of all successors and predecessors respectively. Therefore, using an oracle-based argument similar to the proof in [13] we can show that with respect to a given set of authorization constraints after using our technique to remove hooks, no additional hoisting or removal can be performed, resulting in a minimal placement.

Improving Reuse of Attribute-Based Access Control Policies Using Policy Templates

Maarten Decat, Jasper Moeys, Bert Lagaisse, and Wouter Joosen

iMinds-DistriNet, KU Leuven, 3001 Leuven, Belgium
{first.last}@cs.kuleuven.be

Abstract. Access control is key to limiting the actions of users in an application and attribute-based policy languages such as XACML allow to express a wide range of access rules. As these policy languages become more widely used, policies grow both in size and complexity. Modularity and reuse are key to specifying and managing such policies effectively. Ideally, complex or domain-specific policy patterns are defined once and afterwards instantiated by security experts in their application-specific policies. However, current policy languages such as XACML provide only limited features for modularity and reuse. To address this issue, we introduce policy templates as part of a novel attribute-based policy language called STAPL. Policy templates are policies containing unbound variables that can be specified when instantiating the template in another policy later on. STAPL supports four types of policy templates with increasing complexity and expressiveness. This paper illustrates how these policy templates can be used to define reusable policy patterns and validates that policy templates are an effective means to simplify the specification of large and complex attribute-based policies.

Keywords: Access control, access control policies, attribute-based access control, reuse, modularity, policy templates.

1 Introduction

Access control is an important part of application-level security that constrains the *actions* of authenticated *subjects* on the *resources* in the application by enforcing *access rules*. Traditionally, access control was tightly coupled with the application code, making both hard to maintain. To address this, policy-based access control separates the access rules from the application code into declarative *access control policies* [15]. This approach improves modifiability by allowing the access rules to vary without having to change the application code. This approach also benefits separation of concerns by allowing application developers to focus on writing application code and security experts on specifying the access control policies in separate software modules.

However, while policy-based access control facilitates separation of concerns between application logic and access control logic, the challenge now is to effectively specify and manage the access control policies themselves. For example,

F. Piessens et al. (Eds.): ESSoS 2015, LNCS 8978, pp. 196–210, 2015.

When a subject tries to view the detailed status of a patient:
1. Deny if the subject is not a nurse.
2. Deny if the owning patient has withdrawn consent for the subject , unless the status of the patient is critical or if the subject has triggered breaking−the−glass , which should be logged .
3. Deny if the subject is not on shift .
4. Permit if the subject is currently treating the owning patient .
5. Permit if the subject is of the emergency department .
6. Deny otherwise .

Listing 1.1. A running example of six access control rules from an industrial e-health case study [6]. The application is a patient monitoring system and the rules are specified by the hospital that employs the system. The first rule for which the condition holds should provide the overall decision.

take the access rules of Listing 1.1. These rules are taken from an industrial e-health case study, i.e., a system provided to hospitals for monitoring patients at their homes [6]. These six rules are only an excerpt of the complete set of rules, but already require the concepts of ownership, patient consent, the patient status, breaking-the-glass procedures, the departments of the hospital and the treating relationship between physicians and patients. Moreover, these rules should be combined correctly, in this case being that the first rule for which the condition holds should provide the overall decision. As a result, current applications require a policy language in which a wide spectrum of rules can easily be expressed and combined into one unambiguous composite policy.

Of the current policy languages, XACML [1] partially achieves this. Firstly, by employing attribute-based access control (ABAC, [11]), XACML supports most of the rules required by the case study. Secondly, by employing policy trees (see Figure 1), XACML supports structuring multiple rules and reasoning about how these relate in case of conflict. However, XACML does not allow to specify and manage large policies *effectively*. In the example rules of Listing 1.1, the roles of a subject are a well-known access control concept [7], the concept of patient consent is specific to the domain of e-health, the status of a patient is specific to the patient monitoring system and the departments are specific to the hospital. Ideally, these concepts are defined once as patterns that can be reused within their respective domains. Moreover, each pattern is ideally defined by its respective expert, i.e., an access control expert, an e-health domain expert, the application provider or an expert of the hospital. In terms of attribute-based policies, such patterns would consist of rules and the definitions of the attributes required by these rules. However, XACML does not support attribute definitions and only allows to include other policies without modification such that slight variations cannot be modularized in a single pattern.

To address these limitations, this paper introduces *policy templates* in attribute-based tree-structured policies as part of a novel policy language called

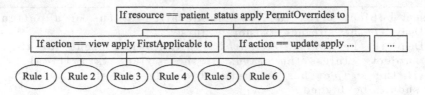

Fig. 1. Example of a policy tree that combines the six rules of the introduction. The leafs of the tree are rules. The intermediate nodes are policies that specify a target and the combination algorithm to combine the results of their children.

the Simple Tree-structured Attribute-based Policy Language or STAPL. Policy templates are policies containing unbound variables that can be bound to values, expressions, rules or even other policies when instantiating the template later on. In addition, STAPL allows to package these templates in reusable *policy modules*. More precisely, STAPL supports four types of policy templates of increasing expressiveness:

1. *simple policy references*, which include other policies without modification,
2. *simple policy templates*, which include variations of policies by extending policy references with unbound variables,
3. *modules of policy templates with attribute definitions*, which encapsulate policy templates and the definitions of the attributes they require,
4. *modules of policy templates with specialized types of attributes*, which extend STAPL with specialized attributes and functions that reason about them.

This paper illustrates how these policy templates and modules can be used to modularize access control patterns and thereby increase policy comprehensibility, facilitate policy reuse and facilitate separation of concerns between the different stakeholders in policy specification.

The rest of this paper is structured as follows. Section 2 further illustrates the problem statement of this work. Section 3 describes STAPL and the different types of policy templates it supports. Section 4 validates the potential of policy templates. Section 5 discusses related work and puts the results in a broader perspective. Section 6 concludes this paper.

2   Context and Problem Illustration

Policy-based access control is an approach to access control in which the access rules are specified in declarative policies that are enforced and evaluated by specialized components in the application. As such, the access rules can vary and evolve separately from the application code and both can be specified by their respective experts, a principle called separation of concerns [14].

Over the last decades, multiple models have been proposed to reason about access control and specify access control policies. The state of the art supports a wide range of complex rules by combining two technologies: attribute-based

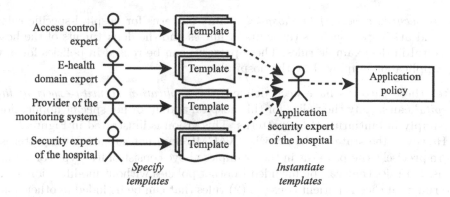

Fig. 2. Our vision on policy specification: an application security expert should be able to specify policies for his or her specific organization and application by instantiating policy patterns represented as templates pre-defined by other experts

access control (ABAC, [11]) and policy trees. Firstly, ABAC is a recent access control model that expresses rules in terms of key-value properties of the subject, the resource and the environment called attributes. Examples of such attributes are the subject roles, the resource location and the time. As such, ABAC supports rules such as identity-based permissions, roles, ownership, time, location, consent and breaking-the-glass procedures. Secondly, the most widely-used language for attribute-based policies XACML [1] additionally allows to combine multiple attribute-based rules in a single policy as a tree, a concept also present in the literature (e.g., [4,12]). As illustrated in Figure 1, each element in such a policy tree defines to which requests it applies and how the results of its children should be combined, e.g., a permit overrides a deny. As said, these technologies together support a wide range of rules. However, specifying complex rules such as the patient consent rule (Rule 2 of Listing 1.1) using these technologies is not trivial. Consequently, managing large sets of such rules is even more challenging.

Modularization and reuse are key to managing this complexity effectively. To illustrate this, again take the example rules from Listing 1.1. Parts of these rules are well-known access control concepts such as the roles of a subject, while other parts are specific to the domain of e-health, specific to the application or specific to the hospital. Ideally, these parts are specified once by their respective experts as patterns that can be reused in multiple policies of their respective domains:

- An *access control expert* defines patterns for well-known access control concepts, such as ownership and hierarchical roles. These patterns can then be reused in any access control policy.
- An *e-health domain expert* defines patterns for frequent e-health rules, such as patient consent and the breaking-the-glass procedure in the example rules. These patterns can then be reused in any e-health policy.
- An *application provider* defines patterns for rules and attributes specific to its application, such as the status of a patient in the example rules. These patterns can be reused in any policy that applies to this application.

- A *security expert of the hospital* defines patterns for hospital-specific rules and attributes, such as the shifts of nurses and the departments of the hospital in the example rules. These patterns can be reused in policies for any application employed by the hospital.

After these patterns have been defined, the *application security expert of the hospital* can specify the policy for this hospital and his or her specific application by simply instantiating these patterns. This vision is illustrated in Figure 2.

However, the state of the art with XACML does not support such patterns. More precisely, the patterns in the examples above consist of five types of definitions: (1) rules that can be included in other policies without modification, e.g., the rule that checks patient consent, (2) rules that can be included in other policies with slight modifications, e.g., the rule that checks whether the subject is on shift, in which the shift hours should be specified by the hospital, (3) attributes that are required by these rules, e.g., the owner of a resource or the current time, (4) attributes that can be used by other rules later on, e.g., the roles of a subject or the status of a patient, and (5) the possible values of a certain attribute, e.g, the departments of the hospital and the roles of the employees in the hospital. Of these five types, XACML only allows the first, i.e., to include other policies literally by using policy references. The more advanced ALFA language for generating XACML policies [8] does support attribute definitions as well, but is still limited to including policies without modification.

To address these limitations, this paper introduces policy templates in attribute-based tree-structured policies as part of a novel policy language called STAPL. Policy templates are partially specified policies that contain unbound variables that can be specified later on. STAPL additionally supports encapsulating these policy templates with the definitions of the attributes they require into self-contained policy modules. Policy templates have been put forward in formal work about policies trees [2] and were part of the Ponder policy language [5], both more than a decade ago. However, policy templates have not been applied in state-of-the-art policy languages of which the complexity actually increases the need for them.

3 Policy Templates in STAPL

This paper introduces policy templates in attribute-based tree-structured policies as part of a policy language called STAPL[1]. More precisely, STAPL supports (1) simple policy references, (2) simple policy templates, (3) modules of policy templates with the definitions of the attributes they require and (4) the extension of the previous with specialized types of attributes. This section first introduces STAPL and then discusses each of these.

STAPL Basics. STAPL is a policy language designed to easily specify attribute-based tree-structured policies. In other words, STAPL takes on a policy model similar to XACML, but is designed for easier specification (amongst

---

[1] The code of STAPL publicly is available at `https://github.com/stapl-dsl/`

```
1    action.id = SimpleAttribute(String)
2    resource.owner = SimpleAttribute(String)
3    subject.treating = ListAttribute(String)
4    subject.roles = ListAttribute(String)
5    environment.now = SimpleAttribute(Time)
6    ...
7    Policy := when (action.id == "view" and resource.type ==
8      "patient_status") apply FirstApplicable to (
9        Rule := deny iff (not "nurse" in subject.roles),
10       Policy := apply PermitOverrides to (
11         Rule := deny iff (subject.id in
12           resource.owner_withdrawn_consents),
13         Rule := permit iff (resource.patient_status == "bad"),
14         Rule := permit iff (subject.triggered_breaking_glass)
15           performing (log(subject.id + " broke the glass"))
16       ),
17       Rule := deny iff (not (environment.now >= 08:00 and
18         environment.now <= 17:00)),
19       Rule := permit iff (resource.owner in subject.treating),
20       Rule := permit iff (subject.department == "emergency"),
21       Rule := deny
22   )
```

Listing 1.2. The STAPL definition of the example rules of Listing 1.1 without the use of policy templates

```
1    // example definitions
2    def defaultDeny = Rule := deny
3    def denyIfNotOnShift = Rule := deny iff (
4      not (environment.now >= 08:00 and environment.now <= 17:00))
5    def permitIfTreating = Rule := permit iff (
6      resource.owner in subject.treating)
7    // example usage
8    action.id = SimpleAttribute(String)
9    ...
10   Policy := when (...) apply FirstApplicable to (
11     ...
12     denyIfNotOnShift,
13     permitIfTreating,
14     ...
15     defaultDeny
16   )
```

Listing 1.3. Example definitions and usage of policy references

others). Listing 1.2 shows the STAPL specification of the example rules of Listing 1.1. As shown, the attributes to be used in the policy are defined first (lines
1-6). These attributes have a type and can be single-valued or multi-valued. Then
the policy is defined in terms of these attributes (lines 7-22). STAPL employs
policy trees consisting of Policies and Rules: Rules are the basic elements of a
STAPL policy and Policies are collections of multiple Rules or other Policies.
Every Policy specifies to which requests it applies by means of an attribute-
based target (following the keyword `when`, see lines 7-8) and specifies how to
combine the results of its children by means of a combination algorithm such as
`PermitOverrides` or `FirstApplicable`. STAPL is defined as an internal domain-
specific language (DSL) in Scala because of its powerful features for both DSLs
and modularity. Of the different features of STAPL, this paper only focuses on
the policy templates, which are explained next.

Simple Policy References.   Policy references are the simplest type of policy
templates provided by STAPL. Similar to XACML, policy references allow policies to be reused without modification and do not include attribute definitions.
Listing 1.3 illustrates how to define a policy reference (lines 2-6) and use one
(lines 10-16). As shown, policy references are Scala functions (denoted by the
keyword `def`) that do not require arguments and return a STAPL Rule or Policy.

Simple Policy Templates.   Policy templates extend policy references with
unbound variables. These unbound variables can be literal values, attribute references, attribute-based expressions or even other Rules or Policies. Listing 1.4
illustrates how to define a policy template (lines 2-9) and use one (lines 11-
18). The first example generalizes `denyIfNotOnShift` of Listing 1.3, the second
encapsulates the pattern of permitting in a certain condition and denying otherwise. As shown, policy templates are Scala functions that require arguments
and return a STAPL Rule or Policy.

Modules of Policy Templates with Attribute Definitions.   Thirdly, STAPL
supports encapsulating policy templates with the definitions of the attributes required by these templates. This decreases the chance for incorrect attribute definitions and fully encapsulates a policy pattern as a reusable and self-contained
module. Listing 1.5 illustrates how to define such a policy module (lines 2-12). As
shown, a policy module is a Scala trait that extends the trait `BasicPolicy`. A Scala
trait is similar to a class, but allows multiple inheritance. The trait `BasicPolicy` defines the variables `subject`, `resource`, `action` and `environment` so that policy modules can assign attributes to them. Listing 1.5 also illustrates how to import the
defined modules, i.e., by extending the scope in which the policies are defined using the Scala traits (line 14). Since Scala allows multiple inheritance using traits,
the scope can be extended with any number of required modules. Moreover, since
Scala allows traits to extend other traits, policy modules can extend existing modules as well. Amongst others, this can be used to express that a certain module
depends on other modules, as illustrated by the `Treating` module (line 9).

```
1   // example definitions
2   def denyIfNotOnShift(start: Time, stop: Time) =
3     Rule := deny iff (not (environment.now >= start
4       and environment.now <= stop))
5   def denyUnless(condition: Expr) =
6     Policy := apply PermitOverrides to (
7       Rule := permit iff (condition),
8       defaultDeny // see Listing 1.3
9     )
10  // example usage
11  action.id = SimpleAttribute(String)
12  ...
13  Policy := when (...) apply PermitOverrides to (
14    ...
15    denyIfNotOnShift(08:00, 17:00),
16    ...
17    denyUnless(subject.department == "emergency")
18  )
```

Listing 1.4. Example definitions and usage of policy templates

```
1   // example definitions
2   trait Shifts extends BasicPolicy {
3     environment.now = SimpleAttribute(Time)
4     def denyIfNotOnShift(...) = ... // see Listing 1.4
5   }
6   trait Ownership extends BasicPolicy {
7     resource.owner = SimpleAttribute(String)
8   }
9   trait Treating extends Ownership {
10    subject.treating = ListAttribute(String)
11    def permitIfTreating = ... // see Listing 1.3
12  }
13  // example usage
14  object example extends BasicPolicy with Shifts with Treating {
15    // notice: no attribute definitions here
16    Policy := when (...) apply PermitOverrides to (
17      ...
18      denyIfNotOnShift(08:00, 17:00),
19      permitIfTreating,
20      ...
21    )
22  }
```

Listing 1.5. Example definitions and usage of policy modules that contain both policy templates as well as the definitions of the attributes they require

```
1   ... // definition of the role attribute omitted
2   // example definition of the role hierarchy
3   val employee = new Role()
4   val nurse = new Role(employee)
5   val headNurse = new Role(nurse)
6   ...
7   // example usage
8   Policy := when (...) apply PermitOverrides to (
9     // this applies to all types of nurses
10    Rule := deny iff (not subject.hasRole(nurse))
11  )
```

Listing 1.6. Example usage of a specialized attribute type for hierarchical roles. The definition of this attribute type required 35 lines of code and is omitted because of space limitations. Notice that `subject` has been extended with a specialized function to reason about hierarchical roles.

Policy Templates with Specialized Attributes. Finally, policy modules can extend the core functionality of STAPL (i.e., single-valued or multi-valued attributes of simple types such as numbers, strings, booleans or dates, and simple operators such as `==`, `in` and `<=`) with specialized types of attributes and functions that reason about these attributes. This functionality can be used to express for example hierarchical roles. In this case, the application security expert would like to define the hierarchy of roles and test whether a subject owns a certain role in this hierarchy. Listing 1.6 shows an example of how such an attribute could be used. For space reasons, Listing 1.6 omitted the definition of this specialized attribute, which consists of the definitions of the attribute itself, the mapping of these attributes to STAPL expressions and the extensions to `subject`, `resource`, `action` and `environment` (35 lines of code in total). This shows that policy modules with specialized attributes are the most expressive type of policy patterns offered by STAPL, but also the most complex.

4   Validation

The previous section discussed the different types of policy templates supported by STAPL. This section validates whether they can be used to create reusable policy modules and that these modules can be used to separate the different stakeholders involved in policy specification.

4.1   Approach

To validate the potential of policy templates, we applied them to an existing policy by consistently factoring out rules and attribute definitions into policy modules. Afterwards, we validated whether each resulting policy module falls within the expertise of a single stakeholder and that each module can be reused within its respective domain. If so, this shows that policy templates can separate the different stakeholders in policy specification, enable reuse and that realistic policies can be specified by instantiating such templates. Of course, this process is mainly meant for validation purposes and STAPL aims for a situation in which policies are specified using policy templates from the start.

The original policy applies to viewing the status of a patient monitored by a patient monitoring system. This policy was specified by the authors before the work on STAPL as part of the e-health case study that was also employed in the running example of Listing 1.1 [6]. The different stakeholders in this policy are the five stakeholders identified before: the general access control expert, the e-health domain expert, the application provider, the security expert of the hospital and the application security expert of the hospital.

The policy was originally specified in XACML and was first translated to STAPL literally to achieve a fair evaluation[2]. In terms of STAPL, the policy consists of 23 Rules (plus 7 default denies) divided over 12 Policies, resulting in a policy tree of depth 5. The policy requires 30 different attributes, of which 17 subject attributes, 11 resource attributes, 1 action attribute and 1 environment attribute.

4.2   Modularization Results

The result of the modularization of the policy is shown in Figure 3. We can make several interesting observations from this figure.

Observation 1: Separation of concerns. Firstly, Figure 3 shows that it is possible to apply policy templates so that every policy module can be specified by exactly one expert. As a result, the different roles of the different experts outlined above can be separated appropriately:

1. The access control expert specifies the general policy modules such as `Roles`, `Time`, `Ownership`, `Location` and `GeneralTemplates`. `GeneralTemplates` amongst others contains the definitions of `defaultDeny` and `denyUnless()` shown in Listing 1.3 and Listing 1.4. `Roles` defines the specialized attribute for hierarchical roles.
2. The domain expert specifies the policy modules that apply to the domain of e-health in general, which in this case is `Consent` and `BreakingGlass`.
3. The application provider specifies the policy modules that are specific to the application, in this case mainly `PatientMonitoringSystem`, which defines the resource attributes supported by the application. Of these attributes, `resource.patient_status` represents a gradation of patient statuses. This specialized attribute has been generalized into the separate module `PatientStatus` that can be reused by the application provider later on.
4. The security expert of the hospital defines the policy modules that are specific to the hospital. These comprise the role hierarchy of the employees of the hospital, its different departments, the definitions of 13 attributes that

---
[2] The original policy and the result of the refactoring are available at https://distrinet.cs.kuleuven.be/software/stapl/

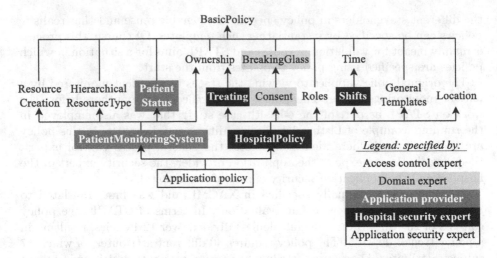

Fig. 3. The hierarchy of policy modules resulting from refactoring the original policy of the e-health case study. Arrows represent a dependency between policy templates and the colors indicate which policy template fits the expertise of which stakeholder.

apply to its employees (e.g., `subject.department` and `subject.is_discharged`) and the rules to reason about shifts and the treating relationship between medical personnel and patients. Again, the latter have been generalized into separate modules that can be reused within the hospital.

5. Finally, the application security expert only specifies the specific policy that holds for the patient monitoring system as used by the hospital. There was no need to define other attributes, i.e., all attributes used by this policy were pre-defined in policy modules.

Observation 2: Reusability. As a second observation, the modules defined by the different experts indeed apply to their own domain and can thus be reused in these domains. For example, the templates in the modules defined by the application provider can be reused by any organization using its application and the templates in the modules defined by the access control expert can be reused in other access control policies regardless of the application or the domain.

Observation 3: Number of policy modules. Thirdly, Figure 3 shows that it was possible to extract no less than 14 modules from a policy consisting of 23 rules. This shows how common reusable patterns are in access control policies and indicates the potential of policy templates for simplifying policy specification. However, we do not expect the number of modules that can be extracted from a larger policy to grow linearly with the number of rules in that policy. On the contrary, given the reusability of the extracted modules discussed in the previous observation, we expect the number of defined modules to stagnate, but the reuse of the templates in these modules to grow.

Observation 4: Fine-Grained Module Hierarchy. Finally, it is worth noting that most of the 14 modules only define a small number of templates and attributes. By employing module dependencies, these fine-grained templates can then be structured in a hierarchy of modules of gradually increasing size in order to only include the specific features needed in the final policy.

4.3   Analysis of the Modules

In addition to the resulting module hierarchy, it is also interesting to analyze the different types of templates used in this hierarchy. More precisely, the 14 modules of Figure 3 define 2 policy references (`defaultDeny` and `defaultPermit`), 6 policy templates (e.g., `DenyUnless()` and `denyIfNotOnShift()`), 30 definitions of standard attributes, 3 definitions of specialized attributes (roles, hierarchical resource types and the status of a patient) and 3 functions for using these specialized attributes. The final policy did not define any other attributes and instantiated 0 policy references, 9 policy templates and 9 functions for using the specialized attributes[3]. These numbers show four interesting observations:

Observation 1: references vs templates. Apart from `defaultDeny` and `default-Permit`, all other templates required unbound variables in order to be reusable. This validates the original claim that policy references by themselves are not powerful enough to provide reusability.

Observation 2: template definition vs instantiation. The number of template instantiations is significantly higher than the number of template definitions. This is mainly due to the use of roles (7x) and `DenyUnless()` (6x) throughout the policy. Both patterns frequently occurred in the policy, while the other templates are more specialized and were used only once. In the former case, modularization provides the benefit of specifying a frequent pattern only once; in the latter case the benefit is having other experts specify complex rules and instantiating these using a simple interface.

Observation 3: no attribute definitions in the application policy. Finally, we observe that all attributes required by the final application policy are defined in the modules and that the policy did not define any attributes itself. This fits reality well: a policy cannot simply define a new attribute since the available resource attributes are determined by the application and defining a new subject attribute requires assigning values for this attribute to the subjects of the organization. In fact, policy modules provide a means to actually consolidate these definitions in both an application-specific or organization-specific module.

4.4   Effect of Templates on the Policy Size

Finally, Table 1 provides metrics about the effect of policy templates on the size of a policy. Table 1 first compares the number of lines of code required to express

---

[3] The final policy did not employ any policy references because `defaultDeny` and `defaultPermit` were encapsulated in `DenyUnless()` and `PermitUnless()`.

Table 1. Metrics about the size of the policy without and with policy templates

	Without	With
#Lines of code for policy definition	57	56 (98.2%)
#Statements in the policy definition	367	205 (55.6%)

the policy (not counting attribute definitions). As shown, the use of policy templates does not significantly affect this number. However, as shown next in the table, the use of policy templates does significantly lower the number of *statements* required to express the policy (e.g., `Policy`, `when`, `==`). This is the result of replacing complex rules or frequently occurring patterns with templates that require less statements to instantiate, such as the consent rule or `DenyUnless()` respectively. While these numbers only comprise one case study, they do show that policy templates have the ability to simplify policy specification significantly.

5   Related Work and Discussion

In the previous sections, we introduced and validated policy templates in attribute-based tree-structured policy languages as a means to improve reuse and modularity. To the best of our knowledge, this work is the first to discuss policy templates in the domain of attribute-based access control. Before this work, policy templates have also been discussed by Wies et al. [17], Casassa Mont et al. [3], Bonatti et al. [2] and as part of the Ponder policy language [5]. Similar to our vision, Wies et al. and Casassa Mont et al. discuss policy templates as part of a larger view on policy management in which templates are a means to allow different experts to cooperate. Bonatti et al. on the other hand discuss policy templates as part of their formal work on modular policy composition using policy trees. As a result, our definition of policy templates aligns to theirs, but fits in the vision of Wies and Casassa Mont. Finally, Ponder is an extensive policy language that amongst others also supports policies with unbound variables, but did not yet employ attribute-based access control or policy trees. The main difference between this work and these four is our focus on attribute-based tree-structured policy templates supported by a concrete policy language, the generalization of policy templates into policy modules and the validation of the potential of policy templates. In addition to these authors, templates have also been discussed in the evolution from role-based access control (RBAC, [7]) to ABAC (e.g., [9,13]). Such role templates were eventually generalized into attribute-based policies, for which STAPL in turn introduces policy templates.

This work started from the observation that state-of-the-art attribute-based policy languages can express a wide range of rules, but that policy specification has also become more complex. This observation aligns to recent visions of amongst others Sandhu [16]. Sandhu argues that attribute-based access control can offer numerous benefits, but also provides risks and challenges such as the increased complexity of attribute management. Similarly, NIST recently published a guide to ABAC [10] containing open challenges and a vision on the multiple roles involved in employing ABAC in an enterprise. This work fits into both visions with a specific focus on efficient policy management and specification. However, when scoping this work in the visions of Sandhu and NIST, it also becomes clear that policy templates are only a first step to an effective deployment of ABAC. For example, while STAPL allows to simply import an attribute definition, the largest effort will be in providing values for this attribute for the appropriate subjects or resources. Therefore, the next step for this work is to extend modular policy management with modular attribute management.

An interesting side-effect of our work on policy modularity is the possibility to explicitly separate the roles of the provider of an application and the organization using it (see Figure 3). In terms of attribute-based policies, the application provider determines the structure of the resources in the application, i.e., the types of resources, their attributes and the actions they support. The organization on the other hand determines the structure of the subjects, i.e., the types of subjects and their attributes. The policies then combine both by expressing which subjects can perform which actions on which resources. As shown in our validation, policy templates provide a means to concretize the structure of the subjects and the resources, which can prove valuable to simplify the correct specification of access rules, for example by employing this information for completeness checking. Towards the future, we plan to further investigate this application of policy templates.

Finally, we want to mention that STAPL is currently built as a Scala DSL. Scala provides powerful features for both DSLs and modularity and allows STAPL to be directly employed in Scala or Java applications. However, in the long run, it is our aspiration that the concepts presented in this paper are incorporated in language-independent policy languages, such as XACML itself.

6   Conclusion

In this paper, we introduced policy templates as a means to improve reuse in attribute-based tree-structured policy languages such as XACML. Our policy language STAPL supports four types of policy templates ranging from simple policy references to modules encapsulating policy templates and the specialized attributes required by these templates. This paper showed that these policy templates can be used to set up a hierarchy of fine-grained reusable policy modules. Each such module can encapsulate a policy pattern, can be defined by the appropriate expert and can be reused within the domain of that expert. As such, it is our aspiration that STAPL is a first step towards a policy specification process in which policy modules are defined once by experts and access control policies are *composed* by instantiating these modules.

References

1. eXtensible Access Control Markup Language (XACML) Version 3.0. OASIS Standard (2013)
2. Bonatti, P., De Capitani di Vimercati, S., Samarati, P.: An algebra for composing access control policies. ACM Trans. Inf. Syst. Secur. 5(1) (February 2002)
3. Casassa Mont, M., Baldwin, A., Goh, C.: Power prototype: towards integrated policy-based management. In: IEEE/IFIP Network Operations and Management Symposium (2000)
4. Crampton, J., Huth, M.: An authorization framework resilient to policy evaluation failures. In: Gritzalis, D., Preneel, B., Theoharidou, M. (eds.) ESORICS 2010. LNCS, vol. 6345, pp. 472–487. Springer, Heidelberg (2010)
5. Damianou, N., Dulay, N., Lupu, E., Sloman, M.: The Ponder policy specification language. IEEE POLICY (2001)
6. Decat, M., Lagaisse, B., Joosen, W.: Middleware for efficient and confidentiality-aware federation of access control policies. Journal of Internet Services and Applications (2014)
7. Ferraiolo, D.F., Sandhu, R., Gavrila, S., Kuhn, D.R., Chandramouli, R.: Proposed NIST standard for role-based access control. TISSEC (2001)
8. Giambiagi, P., Rissanen, E., Nair, S.: Axiomatics Language for Authorization (ALFA). Announced to be Standardized as XACML Profile (April 2014)
9. Giuri, L., Iglio, P.: Role templates for content-based access control. ACM RBAC (1997)
10. Hu, V., Ferraiolo, D., Kuhn, R., Schnitzer, A., Sandlin, K., Miller, R., Scarfone, K.: Guide to Attribute Based Access Control (ABAC) Definition and Considerations. NIST Special Publication (2014)
11. Jin, X., Krishnan, R., Sandhu, R.: A Unified Attribute-Based Access Control Model Covering DAC, MAC and RBAC. In: Cuppens-Boulahia, N., Cuppens, F., Garcia-Alfaro, J. (eds.) DBSec 2012. LNCS, vol. 7371, pp. 41–55. Springer, Heidelberg (2012)
12. Li, N., Wang, Q., Qardaji, W., Bertino, E., Rao, P., Lobo, J., Lin, D.: Access control policy combining: Theory meets practice. ACM SACMAT (2009)
13. Li, N., Mitchell, J.C., Winsborough, W.H.: Design of a role-based trust-management framework. IEEE Security and Privacy (2002)
14. Parnas, D.L.: On the criteria to be used in decomposing systems into modules. Communications of the ACM 15(12), 1053–1058 (1972)
15. Samarati, P., de Capitani di Vimercati, S.: Access control: Policies, models, and mechanisms. In: Focardi, R., Gorrieri, R. (eds.) FOSAD 2000. LNCS, vol. 2171, p. 137. Springer, Heidelberg (2001)
16. Sandhu, R.: The authorization leap from rights to attributes: Maturation or chaos? In: Proceedings of the 17th ACM Symposium on Access Control Models and Technologies, SACMAT 2012. ACM (2012)
17. Wies, R.: Using a classification of management policies for policy specification and policy transformation. In: Integrated Network Management IV, pp. 44–56. Springer (1995)

Monitoring Database Access Constraints with an RBAC Metamodel: A Feasibility Study

Lars Hamann, Karsten Sohr, and Martin Gogolla

University of Bremen, Computer Science Department
D-28334 Bremen, Germany
{lhamann,sohr,gogolla}@informatik.uni-bremen.de

Abstract. Role-based access control (RBAC) is widely used in organizations for access management. While basic RBAC concepts are present in modern systems, such as operating systems or database management systems, more advanced concepts like history-based separation of duty are not. In this work, we present an approach that validates advanced organizational RBAC policies using a model-based approach against the technical realization applied within a database. This allows a security officer to examine the correct implementation – possibly across multiple applications – of more powerful policies on the database level. We achieve this by monitoring the current state of a database in a UML/OCL validation tool. We assess the applicability of the approach by a non-trivial feasibility study.

Keywords: Model checking for security, Models for security, Verification techniques for security properties, Security by design.

1 Introduction

Modeling systems with languages like UML [14] and OCL [15] offers many advantages for the development process: Models allow developers to state, analyze and predict interesting characteristics of the system under study before an actual implementation is done, and models allow developers to specify the implementation of the intended system.

RBAC (Role-Based Access Control) is a well accepted approach for designing and implementing access management. Typically, many proposed approaches use RBAC in the system design phase in a forward engineering way. RBAC, however, also allows us to monitor access violations in a running system. Monitoring approaches can be employed for existing systems during running operations.

This contribution puts forward an RBAC modeling approach and is based on previous foundational work on an RBAC metamodel [11] and on runtime monitoring using UML and OCL [10]. This new approach combines both lines of work by integrating them and evaluating the applicability using relational database systems. We concentrate on databases rather than the application itself as access violations might also occur at this lower system level; the lower layer might not respect the policies defined for the application level [2]. The evaluation is done by a feasibility study on a publicly available moderate sized database [20].

F. Piessens et al. (Eds.): ESSoS 2015, LNCS 8978, pp. 211–226, 2015.

The rest of this paper is structured as follows. Section 2 introduces the essential RBAC concepts needed in our context and explains a running example. Section 3 sketches the UML and OCL tool USE (UML-based Specification Environment) and its monitoring features. Section 4 explains the details of monitoring RBAC on a relational database. Special emphasis is laid on dynamic separation of duty constraints. Section 5 discusses a feasibility study for our approach and shows that it works in a case study where some 100,000 database tuples are monitored and some 10,000 access operations on the tuples and violation of dynamic separation of duty rules are analyzed. Section 6 discusses related work. Section 7 summarizes the results and gives an outlook on future work.

2   Role-Based Access Control

RBAC is widely used in organizations such as financial institutes or enterprises for access management. Users do not obtain permissions to access resources directly, but through roles. Roles often correspond to job functions that a user holds within her organization. The role concepts have been described in the RBAC ANSI standard [1]. RBAC comprises the sets *Users*, *Permissions*, and *Roles*, as well as the relations *UA*, *PA*, and *RH*. *UA* is a many-to-many relation which represents the roles assigned to users ("user assignment"). The assignment of permissions to roles is expressed by the many-to-many relation *PA*. Furthermore, permissions are often seen as pairs of resources and actions, e. g., the action "approve" is allowed to be performed on the resource "cheque".[1] *RH* describes a role hierarchy relation on the set of *Roles*, i.e., roles can inherit permissions from other roles.

The aforementioned RBAC sets and relations are shown in Fig. 1 where we present a UML-based metamodel of RBAC. For example, the *Users* and *Roles* sets are represented by the `User` and `Role` classes; the *UA* relation is expressed by the association between both classes. Please note that permissions are represented by an association class between the `resource` and `action` classes.

RBAC also supports advanced access control concepts, such as role hierarchies and role-based authorization constraints. The role hierarchies are represented by a senior/junior association from the `Role` class to itself in Fig. 1. A typical example of a role hierarchy relation is given by the roles `Cashier` and `Cashier Supervisor` where the former role is junior to the latter.

Authorization constraints allow a security officer to express organizational rules. The most well-known kind of authorization constraint is separation of duty (SoD). SoD prevents a user from committing fraud by splitting tasks into several parts, which must be executed by different users [18]. Two forms of SoD are usually distinguished, namely, static and dynamic SoD. Static SoD

---

[1] In the literature on access control, usually the terms "object" and "operation" are used instead of "resource" and "action". However, we felt that the former terms could be mixed with the notions of object and operation in UML.

is often expressed in terms of mutually exclusive roles, e.g., the roles `Clerk` and `Supervisor`. Static SoD is enforced at administration time, i.e., when a user is assigned to a role. In contrast, dynamic SoD is enforced at runtime. For example, a banking clerk who has approved a cheque is not allowed to validate it. Such conditions often need the access history for access decisions [18]. As a distinguishing feature, the RBAC metamodel given in Fig. 1 also supports this kind of dynamic SoD, which is represented by the class `DynamicSoD`.

Figure 2 visualizes our concept of modeling dynamic SoD and is meant for didactic purposes. After a user has applied the action `preAct` to the resource property `preProp` belonging to a resource of type `preT`, she is not permitted to apply `postAct` to the resource property `postProp` of a resource of type `postT` (dynamic SoD constraint). In the aforementioned example, the resource types `preT` and `postT` equal "cheque". The preaction is "approve", whereas the postaction is "validate". The properties are then "approved" and "validated", respectively.

3 USE

3.1 Validation and Verification with USE

Modeling features and their analysis through validation and verification is supported by the tool USE (UML-based Specification Environment) [7]. Within USE, UML class, object, statechart, sequence, and communication diagrams extended with OCL are available. USE assists the developer in order to validate and verify model characteristics. Validation and verification can be realized in USE by employing a model validator based on relational logic and SMT solvers. Model properties to be inspected include consistency, redundancy freeness, checking consequences from stated constraints, and reachability. These properties are handled on the conceptual modeling level, not on an implementation level. Employing these instruments, central and crucial model characteristics can be successfully and efficiently analyzed and checked.

3.2 Monitoring with USE

The USE Monitor project was started as a USE plug-in to support runtime verification of Java applications [10]. The monitor allows a developer to attach to a running application, take a snapshot of its current state and validate this snapshot against defined constraints. Using the monitor, an application can be verified at a more abstract level than the code, because the used model can be a small part of the overall system, by dropping unnecessary details. For example, a huge inheritance hierarchy can be compressed to those super classes required for the validation task. After an initial snapshot has been taken, the suspended application can be resumed to monitor changes in the system. While listening to change events, like the creation of new instances or operation calls, constraints defined in the model are validated, e.g., when the monitor receives an operation call event to an operation that is considered in the model, it evaluates the

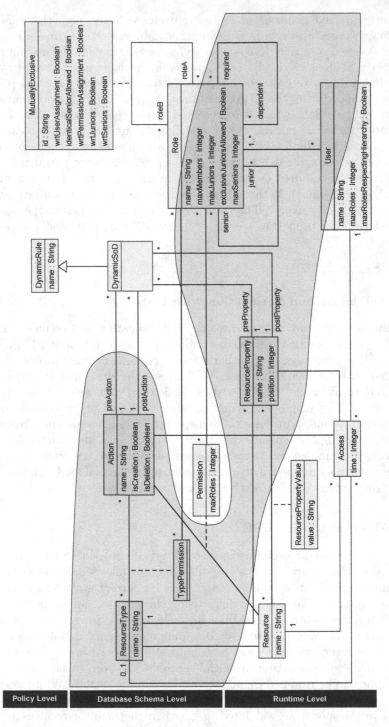

Fig. 1. Excerpt of the RBAC metamodel

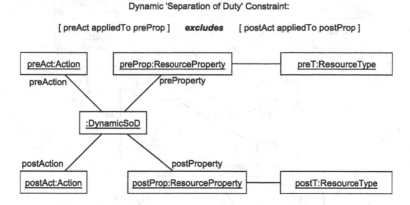

Fig. 2. Realizing dynamic SoD in our RBAC metamodel

defined preconditions and, if present, possible state machine transitions for this operation. The user of the monitor is informed immediately about any unexpected behavior of the application, i. e., about model constraint violation. While the first version of the monitor was limited to the Java runtime environment, the current version allows for an easy integration of other target platforms. This is achieved by using so-called adapters. These adapters detach the monitor from a concrete target platform by hiding the concrete communication between the target platform and the monitor. Moreover, an adapter can map non object-oriented platforms (i. e., relational databases) to the object-oriented world of UML and OCL as it is done in this work.

4 Monitoring RBAC Constraints with UML and OCL Models

This section describes our approach of verifying RBAC policies defined on an abstract level against low level implementations in a step-wise manner. Figure 3 shows an overall picture of it. A central part of this approach is the RBAC metamodel described in detail in [11].

An excerpt of a slightly modified and extended version of this metamodel can be seen in Fig. 1. A central extension was done by integrating type information into the metamodel by adding the classes `ResourceType` and `ResourceProperty`. The RBAC specification is very general about the concrete meaning of a resource, but for our approach we need to distinguish between the permissions on a type, in our example a table, and the permissions on access to an instance, i. e. a table entry, of these types.

Before we discuss the steps in detail, we briefly describe the overall workflow. In Step 1, the USE monitor is utilized to retrieve any information that is relevant to the overall RBAC policies from a database. Relevant information include defined roles, tables, columns and database users. These elements are created within the USE tool as objects according to the RBAC metamodel.

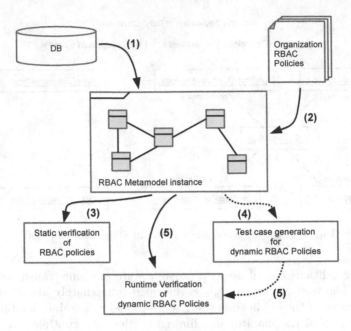

Step 1: Retrieve RBAC relevant information from database
Step 2: Apply organization RBAC policies
Step 3: Validate DB permissions against RBAC policies
Step 4: Generate test cases for dynamic aspects
Step 5: Runtime verification of dynamic RBAC policies

Fig. 3. Overview of the verification process

After this, the developer applies the organization RBAC policies to the information extracted from the database in Step 2. These organization policies contain rules that cannot be specified in the database. Examples are static and dynamic SoD constraints, which cannot be expressed in most database systems. Simple mismatches between the organization policies and the database security configuration, like missing roles or resources can already be identified in this phase of the process.

In Step 3, basic and extended verification jobs of the organization policies can be run. The basic verification evaluates the defined invariants on the metamodel in order to discover violations of static RBAC constraints, e.g., the violation of mutually exclusive roles defined in the organization policies. Extended verification applies more powerful model finding techniques to verify assumptions about the configuration, e.g., whether a given workflow is executable with the current security settings.

The same model finding approach is used in the optional step 4 to generate test cases for validating dynamic RBAC policies. It can be used to generate scenarios (workflows) that violate a given dynamic policy. These scenarios can be used to test if the applications using the database take into account the organizational dynamic RBAC policies.

For Step 5 the USE monitor is utilized to listen to database changes. After interesting change events, USE can validate the organizational dynamic RBAC policies at runtime and report violations to the developer. This step allows an application-independent runtime verification of dynamic policies. The test cases generated in Step 4 can be used as a test driver or the monitoring is done during "ordinary work", if the performance impact of the monitoring is not an issue.

After this summary of the overall process, we now discuss each step in detail.

4.1   Retrieve RBAC Relevant Information from a Database

To be able to verify more general security settings, i. e., RBAC policies, against the concrete permissions present in the database, the concrete permissions need to be available to a verification tool. For our approach we developed an SQL RBAC adapter that reads the database schema, the defined permissions and – if required – the current database state. Table 1 shows the currently used mappings from relational database elements to the metaclasses of our RBAC metamodel.

Table 1. Element mapping from Database elements to RBAC metamodel classes

Element	Catalog Item	$\rightarrow$ RBAC metamodel class
Table	information_schema.tables	$\rightarrow$ ResourceType
Column	information_schema.columns	$\rightarrow$ ResourceProperty
User*	pg_user	$\rightarrow$ User
Role*	pg_group	$\rightarrow$ Role
Permission Kind	Fixed values (INSERT, ...)	$\rightarrow$ Action
Table Permission	information_schema.table_privileges	$\rightarrow$ TypePermission
Tuple*		$\rightarrow$ Resource
Value*		$\rightarrow$ ResourcePropertyValue

*These elements require special treatment and are described in Sect. 4.1.

This table is just a brief overview to illustrate the idea of the mapping. It does not contain all relevant information. For example, we do not provide information about the role membership and role hierarchy. For this kind of relation, the mapping is typically done by using foreign key information of the corresponding relations. The information for the metaclasses ResourceType, ResourceProperty and TypePermission can be retrieved by querying the database information schema as it is defined in the SQL standard (c. f. [8]). Each returned tuple is considered as a new instance of the mapped metaclass. Figure 5 shows the object diagram that is extracted by the monitor after it was applied to a database containing the table shown in Fig. 4.

In the upper part of this figure, the table structure can be seen. Directly below the structure of the table, its content is shown. In the example the content of the

cheque	nr [PK] text	amount numeric(10,2)	approved integer	validated integer
1	100005	150.00	-1	0
2	100006	120.00	-1	-1

Fig. 4. Table *cheque* with sample data

table consists of two entries. The cheque with the number 100005 and `amount` 150.00 has been approved (the property `approved` $\neq 0$) but not validated (the property `validated` $= 0$). The other cheque with number 100006 about 120.00 has already been validated and approved. The values of a table can be extracted by taking into account the information about the table schema and constructing corresponding SQL statements. Since not all data contained in a database are required for the following tasks, adequate filters could be used to reduce the overall size of the resulting object diagram. Specifically, the applicability of model finding during some tasks depends on this reduction.

The permissions on resource types, i.e., tables, can be retrieved by querying the default information scheme, too. However, the concrete users and roles are not easy to query using the SQL standard. In our work, we used the views `pg_roles` and `pg_users` which are specific to the database system PostgreSQL[2]. Since the possible actions on tables are defined in the SQL standard, they can be defined beforehand. In our example, the defined permissions (instances of association class `ResourcePermission`) and their assignments to roles (represented as links) can be seen on the left side of Fig. 5. One can retrieve that the role `Supervisor` has the following permissions on table `cheque`: DELETE (object `tp3`), UPDATE (`tp2`), and SELECT (`tp4`), whereas, the role `Clerk` has the permissions SELECT (`tp4`), INSERT (`tp1`), and UPDATE (`tp2`).

4.2 Apply Organizational RBAC Policies

In this step, the developer enriches the monitored information with RBAC policies from the organization that cannot be expressed in the used database system. For example, the RBAC concept of dynamic SoD with respect to access history is unsupported in database systems.

To apply these more general policies, the developer needs to modify the instance of our RBAC metamodel read in Step 1. She can set attribute values of already present metamodel elements, such as roles, define new rules by creating rule objects and, link instances to fit the organization needs. Please note that a developer only needs to configure instances of provided rules. She does not need to write policy rules in OCL, since these rules are already defined in the metamodel. Further, we only show one example of a policy (DynamicSoD), but other policies can be integrated easily by a metamodel designer into the RBAC

---

[2] Using Microsoft SQL Server the corresponding system view is `syslogins`. The column `issqlrole` determines if an entry is a role or a user.

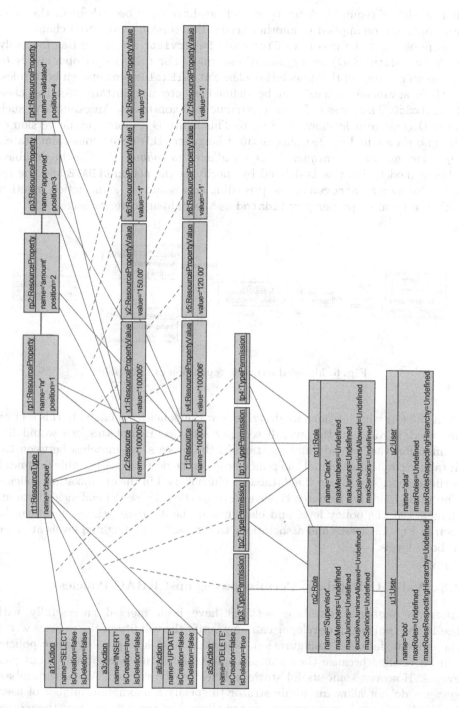

Fig. 5. The table shown in Fig. 4 as an instance of the RBAC metamodel, generated with the USE tool

metamodel, if required. After these policies have once been defined, they can automatically be applied in another verification session by our tool chain.

Suppose that the two roles `Clerk` and `Supervisor` are defined as mutually exclusive (static SoD) by organizational rules. For this, a developer needs to create an instance of the association class `MutualExclusive` between both roles.

More advanced policies can be defined by creating instances of the class `DynamicSoD`. This class enforces a dynamic SoD constraint. An example of such a dSoD constraint is shown in Fig. 6. This figure is a more specific version of the one shown in Fig. 2. Using natural language, this dSoD constraint states: *"If a user approves a cheque, she is not allowed to validate it."* For the database schema model, this rule is defined by specifying the action `UPDATE` and the resource property `approved` as the preceding access and the same action together with the resource property `validated` as a forbidden posterior access.

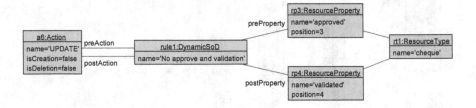

Fig. 6. Modeled dynamic separation of duty policy

In the following sections we describe how the policies explained in this section are used to validate the overall security settings. Before this, we would like to mention that already at this stage in the process, mismatches between the database configuration and the policy level can be detected. A possible mismatch is the absence of a role in the database, which is used in the organization policies. The same holds for resource types and properties. These dependencies between elements at the policy level and elements at the database schema level can be seen in Fig. 1. Classes and associations that cross the highlighted levels are used in both levels.

4.3   Validate Database Permissions against RBAC Policies

After the concrete database settings have been merged successfully with the RBAC policies, a developer can now verify the validity of the policies w. r. t. the current database configuration. Note that in this step no dynamic policies can be verified because the monitored data contain no information about past events. However, some useful static policies can be verified. As most database systems do not allow an administrator to specify a maximum number of users for a role, e. g., only two supervisors are allowed in a given company, these organization policies can now be verified. Our RBAC metamodel contains invariants specified in OCL for several extended RBAC policy types, including the maximum number of roles a user can be a member of, mutually exclusive roles and role hierarchies.

Using our validation and verification tool USE, violations of these invariants can be discovered. USE evaluates the given invariants against the system state which was created in the previous step. If a constraint fails, USE provides a rich set of functionality to discover the violating elements. The outcome of this phase can be used to give advice to database administrators how to change the security settings within the examined database.

Another question which can be answered in this step is if a given workflow is executable w.r.t. current database permissions. For our example, we want to know, if a cheque can be approved and afterwards be validated. For this small example a database administrator might directly see that the workflow is executable. However, for more complex workflows automatic verification techniques are needed. Our toolchain supports this task by using the model validator plug-in described in Sect. 3.2. To do so, a developer needs to specify the workflow to validate by means of declarative assumptions that need to be fulfilled. The workflow to validate a cheque requires to describe a correct execution, i.e., it has to be specified that a cheque is approved and validated. This can be done by defining an invariant that enforces two access actions to the same resource of type named *cheque*: one for the update of the resource property `approved` and one for the update of the property `validated`. This invariant together with other settings, like bounds for the number of instances for each type, are provided to the model validator, which then tries to find a valid object diagram w.r.t. the default RBAC constraints and the additional invariant. The model validator uses the previously created object diagram as a starting point, a so-called partial solution. If the model validator finds a solution within the provided bounds, the workflow is executable under the current permissions of the database. If it does not find a solution, one cannot directly state that the workflow is not executable, because the search space of the model validator is limited by the provided bounds. This means, a valid solution might exist outside the configured bounds. A developer can now increase the bounds until it is likely that no valid solution exists. Hints to the developer why a given setup is not satisfiable can also be provided by the model validator.

4.4   Generate Test Cases for Dynamic Aspects

The model validator can be used for test case generation, too. For this task, the approach described in the previous section is slightly changed: instead of providing information about a valid workflow, the model validator is now used to find a solution that does not fulfill a given constraint, e.g., the invariant or invariants that define a dynamic SoD policy. This negation of invariants is directly supported by the model validator, since it is useful for several examination tasks in the context of model finding.

Figure 7 shows a resulting test case using the cheque example. The model validator extended the object diagram shown in Fig. 5 by several instances.

Note that we do not show all instances again. Only the relevant ones for the invalid workflow are displayed. All other are hidden in the object diagram, but are present in the used system state. In detail, the model validator added three access objects each representing an access to a resource or resource property of the resource type `cheque` by the user `bob`. The first access (`time=2`) creates resource `r1` using the INSERT action (`a3`). Afterwards, the access sequence violating the dynamic SoD rule shown in Fig. 6, which updates the properties `approved` (at time 28) and `validated` (at time 32) occur. These generated test cases can be seen as basic execution sequences that violate a given high-level policy. The concrete workflow execution, possibly using different applications, depends on the application landscape present. Therefore, a security officer needs to map the generated basic access operations to a business workflow to execute the test case.

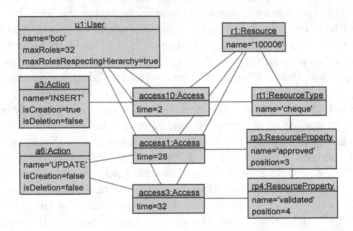

Fig. 7. Generated test case for an invalid dSoD policy

4.5   Runtime Verification of Dynamic RBAC Policies

Until now, we only considered the database state as it is. However, using the monitor plug-in dynamic verification could be applied, too. For this, the database management system needs to provide a notification mechanism that allows an application to register for interesting events. These events include statements for the actions INSERT, UPDATE, and DELETE. An example of such mechanism are the Microsoft Notification Services [12] provided by SQL Server 2005. Starting with SQL Server 2008 these services are integrated into the SQL Server Reporting Services [13]. Using PostgreSQL a notification infrastructure can, for example, be built by using the non-standard SQL command NOTIFY [19]. Both mechanisms have in common that the notifications are sent asynchronously. Since our approach is designed to identify policy violations and not to prevent them, this asynchronous behaviour fits well. Only the sequence of access must be correct.

In addition, using asynchronous notifications reduces the overhead for the running applications, since the validation can be done independently.

Using such an appropriately configured infrastructure, the monitor could react on events by notifying the USE tool about changes. USE could then validate all defined RBAC policy rules based on its knowledge about the history of executed commands. This history would be built incrementally starting at the first received command. Commands executed before cannot be considered. Therefore, to verify a complete workflow, this workflow needs to be executed as a whole during the runtime verification task. One benefit of this runtime verification approach would be that the correct implementation of RBAC policies within a single application or across multiple applications using the same data source can be checked, because the evaluation of the policies is done on the database layer.

5 Feasibility Study

Until now, we only showed small examples adequate for presenting the overall ideas and explained concepts of a possible runtime verification task. To get an early impression about the database sizes that our approach can handle, we extracted database information of various sizes and generated 10,000 access operations. Afterward, we ran a complete validation of the structure, i.e., we checked the extracted object diagram against the multiplicity constraints and the (OCL) invariants present in our metamodel. Further, we ran these validations using different numbers of RBAC policy rules. Starting with seven rules and ending with 28. Table 2 summarizes our results. All performance evaluations, except the second row, were executed on a workstation powered by a Pentium DualCore E5300 with 2.6GHz with 4GB RAM running Ubuntu 14.04LTS. USE was executed using Oracle JRE 1.7 with an allowed heap memory size of 2GB. The results in the second row of the table are collected from a notebook with a similar configuration, but running Windows 7 32bit, which only allows a maximum Java heap size of some 1GB. The complete RBAC metamodel including detailed information about the evaluation can be found online [9].

Table 2. Performance evaluation

				Structure				Invariants			
#Rec.	#Inst.	#Links	read	7	14	21	28	7	14	21	28
74,964	638,897	75,009	52s	1s	1s	1s	1s	4.7s	8.8s	12.1s	15.5s
(Windows 7)			(53s)	(1.6s)	(1.6s)	(1.6s)	(1.6s)	(6.4s)	(11.5s)	(17.4s)	(23.1s)
89,222	769,329	89,267	60s	1.4s	1.4s	1.3s	1,2s	5s	9.6s	12.7s	16.8s
112,366	986,206	112,409	78s	1.6s	1.6s	1.5s	1.5s	4.7s	8.7s	12.8s	17.1s

Validation time with n policy rules

In summary, we can state that our approach is able to validate RBAC policies in about 20 seconds on a moderate sized database (with about 100,000 tuples taking into account 10,000 access operations) and including on the RBAC side about 30 dynamic SoD rules (following the template in Fig. 2 and instantiated in example form in Fig. 6). Given the presented results, one can see that both the creation of the database state (column *read*) and the validation of the invariants grow linear. Further, the measured durations are still acceptable when taking into account that they represent a validation of the whole system state.

Currently, the memory consumption of USE in combination with the RBAC monitor is quite large. Using 1GB heap space, only the first database state containing roughly 75,000 records is manageable. This number can be increased, since the RBAC monitor currently keeps a copy of all read database rows for performance reasons. USE as a standalone application or used as a library can still handle more instances.

6   Related Work

Our approach builds on an RBAC metamodel [11] that we have extended with the notions of ResourceType and ResourceProperty (see Section 4). Several other works exist that model RBAC policies with UML and OCL [4,3,17,6]. Some of those approaches also target at validating RBAC policies, e.g., [3]. However, none of those works deal with runtime monitoring and they do not support dynamic SoD constraints based on the access history.

Only simple SoD concepts have been implemented in database management systems (DBMS) until now, such as mutually exclusive roles w.r.t. role membership (static SoD) or role activation (simple dynamic SoD). Advanced role-based authorization constraints, such as history-based dynamic SoD, are not supported. RBAC concepts for relational databases have also not been discussed much in scientific literature in recent years. In older work, Ramaswamy and Sandhu discuss RBAC concepts that are supported by commercial DBMS [16]. Later, Bertino and Sandhu came to the conclusion that commercial DBMS use only limited RBAC concepts [5]. Our experience with current versions of DBMS still supports this statement as mentioned before.

7   Conclusion

We have employed a UML and OCL model monitor for RBAC in relational databases. The approach can detect, for example, the violation of dynamic mutually exclusive roles specified as organizational policies. We have performed a feasibility study where a moderate relational database with some 100,000 tuples together with about 10,000 database access operations have been monitored in our tool USE. The approach allows one to check the consistency of the specified policies with actual workflows. We have concentrated on dynamic SoD constraints with respect to an access history.

The current approach is only a first step towards monitoring RBAC policies during runtime. Future work should study further types of RBAC constraints which can be captured by our metamodel. Currently, the formulation of the organizational policies in form of object diagrams is not user-friendly. A better syntactical presentation should be developed. The utilization of the model validator and its features could also be improved, and further steps in the monitoring process, like schema and constraint extraction, could be automatized.

References

1. American National Standards Institute Inc.: Role Based Access Control, ANSI-INCITS 359-2004 (2004)
2. Anderson, R.J.: Security Engineering: A Guide to Building Dependable Distributed Systems, 2nd edn. Wiley Publishing (2008)
3. Basin, D.A., Clavel, M., Doser, J., Egea, M.: Automated analysis of security-design models. Information & Software Technology 51(5), 815–831 (2009)
4. Basin, D.A., Doser, J., Lodderstedt, T.: Model driven security: From UML models to access control infrastructures. ACM Trans. Softw. Eng. Methodology 15(1), 39–91 (2006)
5. Bertino, E., Sandhu, R.: Database Security-Concepts, Approaches, and Challenges. IEEE Trans. Dependable Secur. Comput. 2(1), 2–19 (2005)
6. Fernández-Medina, E., Piattini, M.: Extending OCL for secure database development. In: Baar, T., Strohmeier, A., Moreira, A., Mellor, S.J. (eds.) UML 2004. LNCS, vol. 3273, pp. 380–394. Springer, Heidelberg (2004)
7. Gogolla, M., Büttner, F., Richters, M.: USE: A UML-Based Specification Environment for Validating UML and OCL. Sci. of Comp. Prog. 69, 27–34 (2007)
8. Gulutzan, P., Pelzer, T.: SQL-99 complete, Really – An Example-Based Reference Manual of the New Standard. R&D Books (1999)
9. Hamann, L., Gogolla, M., Sohr, K.: RBAC meta-model and detailed evaluation results,
   http://www.db.informatik.uni-bremen.de/publications/intern/
   RBACEvaluation.use (last visited: March 30, 2014)
10. Hamann, L., Hofrichter, O., Gogolla, M.: OCL-Based Runtime Monitoring of Applications with Protocol State Machines. In: Vallecillo, A., Tolvanen, J.-P., Kindler, E., Störrle, H., Kolovos, D. (eds.) ECMFA 2012. LNCS, vol. 7349, pp. 384–399. Springer, Heidelberg (2012)
11. Kuhlmann, M., Sohr, K., Gogolla, M.: Comprehensive Two-Level Analysis of Static and Dynamic RBAC Constraints with UML and OCL. In: Proc. Secure Software Integration and Reliability Improvement (SSIRI 2011), pp. 108–117. IEEE (2011)
12. Microsoft: SQL Server Notification Services,
   http://technet.microsoft.com/en-us/library/ms172483%28v=sql.90%29.aspx
   (last visited: February 05, 2014)
13. Microsoft: SQL Server Reporting Services (SSRS),
   http://technet.microsoft.com/en-us/library/ms159106.aspx (last visited: February 05, 2014)
14. UML Superstructure 2.4.1. Object Management Group (OMG) (August 2011),
   http://www.omg.org/spec/UML/2.4.1/Superstructure/PDF
15. Object Constraint Language 2.3.1. Object Management Group (OMG) (January 2012), http://www.omg.org/spec/OCL/2.3.1/

16. Ramaswamy, C., Sandhu, R.: Role-Based Access Control Features in Commercial Database Management Systems. In: Proc. of 21st National Information Systems Security Conference, pp. 503–511 (1998)
17. Ray, I., Li, N., France, R.B., Kim, D.K.: Using UML to visualize role-based access control constraints. In: Proc. of the 9th ACM Symp. on Access Control Models and Technologies, pp. 115–124 (2004)
18. Simon, R., Zurko, M.: Separation of duty in role-based environments. In: 10th IEEE Computer Security Foundations Workshop (CSFW 1997), pp. 183–194 (1997)
19. The PostgreSQL Global Development Group: PostgreSQL 9.3.2 Documentation: NOTIFY, `http://www.postgresql.org/docs/9.3/static/sql-notify.html`, (last visited: February 05, 2014)
20. Treat, R., Mohan, V.: pgFoundry: Sample Databases, dellstore2, `http://pgfoundry.org/projects/dbsamples/` (last visited: March 20, 2014)

Author Index